KV-510-901

NATURAL ENVIRONMENT RESEARCH COUNCIL
INSTITUTE OF TERRESTRIAL ECOLOGY

METALS IN ANIMALS

ITE SYMPOSIUM NO. 12

EDITED BY

D. OSBORN

ITE, MONKS WOOD EXPERIMENTAL STATION,
ABBOTS RIPTON, HUNTINGDON

PROCEEDINGS OF A WORKSHOP HELD AT MONKS WOOD EXPERIMENTAL STATION
15-16 SEPTEMBER 1982

Printed in Great Britian by NERC/SERC Reprographic Services, Swindon

C NERC Copyright 1984

Published in 1984 by
Institute of Terrestrial Ecology
68 Hills Road
Cambridge
CB2 1LA

ISBN 0 904282 77 5
ISSN 0263 8614

LIVERPOOL INSTITUTE
OF HIGHER EDUCATION
Accession No.
130212
Class No.
Q 591.1 OSB
Catal.
8/10/90

ACKNOWLEDGEMENTS

Mrs June Stokes prepared the camera-ready typescript from the author's
original material. Her diligence in doing this exacting job is greatly
appreciated. Thanks are also due to John Williamson for drawing several
of the figures, to Mrs Jean King and John Beckett of the ITE library for
advice on references, and to Penny Ward for editorial assistance.

COVER ILLUSTRATIONS

Sketch studies of crab by C F Tunnicliffe OBE RA, by kind permission
of the beneficiaries of his estate
Bank vole photograph Dr P A Morris, Royal Holloway College, London
Fulmar photograph Dr R K Murton
Puffin photograph Dr M P Harris
Swan on River Thames photograph Dr M Birkhead, Oxford University
Cover design C B Benefield

The **Institute of Terrestrial Ecology (ITE)** was established in 1973,
from the former Nature Conservancy's research stations and staff,
joined later by the Institute of Tree Biology and the Culture Centre
of Algae and Protozoa. ITE contributes to, and draws upon, the
collective knowledge of the fourteen sister institutes which make up
the **Natural Environment Research Council,** spanning all the
environmental sciences.

The Institute studies the factors determining the structure,
composition and processes of land and freshwater systems, and of
individual plant and animal species. It is developing a sounder
scientific basis for predicting and modelling environmental trends
arising from natural or man-made change. The results of this research
are available to those responsible for the protection, management and
wise use of our natural resources.

One quarter of ITE's work is research commissioned by customers, such
as the Department of Environment, the European Economic Community, the
Nature Conservancy Council and the Overseas Development
Administration. The remainder is fundamental research supported by
NERC.

ITE's expertise is widely used by international organizations in
overseas projects and programmes of research.

D Osborn
Institute of Terrestrial Ecology
Monks Wood Experimental Station
Abbots Ripton
HUNTINGDON
Cambridgeshire
PE17 2LS

04873 (Abbots Ripton) 381

CONTENTS

Page No.

L.I.F.E.
THE ... LAND LIBRARY
... RD., LIVERPOOL, L16 8..

INTRODUCTION

Toxic metals are part of the environment. Man's industry and agriculture redistribute these metals, concentrating them from ores, discarding others in effluents, and even applying some to crops as pesticides. In some parts of the world, substantial leaching of metals from soils may be occurring because of groundwater acidification. So, as a result of man's activities, animals and plants can be exposed to abnormal amounts of toxic metals. This exposure can lead to the accumulation of pharmacologically significant concentrations of metals in plant and animal tissues. Interestingly, there are also circumstances in which plants and animals accumulate toxic metals from natural sources.

Clearly, the environmental and ecological significance of these metal burdens in man, agricultural products and wildlife needs to be assessed, and suitable measures taken to minimise any harm that is being done.

For this reason, much research into the effects of toxic metals has been done in recent years, and there are now many examples of adverse effects on wildlife, domesticated animals, and man. However, when working with toxic metals, the essential role that many metals play in normal physiology must never be forgotten. Studies of toxic metals, whether done for environmental monitoring or to determine harmful effects, will almost always be improved by simultaneous consideration of the essential metal status of the animal, plant or ecosystem concerned. This can be illustrated in a number of ways: for example, the amount of essential metal in the diet, the amount in any one tissue or biochemical compartment, can influence the absorption and distribution of the toxic metal and hence, in all probability, modulate its effects.

Furthermore, in common with studies on other toxic chemicals, understanding the effects of a toxic metal depends on a thorough knowledge of the affected species' normal physiology. Such understanding is particularly necessary if we wish to make an attempt to assess the importance of an effect of a toxic metal in the life of wild animals or plants, because, while a toxic metal may disrupt a particular biochemical or physiological function in all animals or plants, the importance of that function will vary in different animals and plants and may even vary from time to time in any one species. Changes in functional response could lead to widely different whole animal manifestations of the effects of metals, and suggest that we cannot make environmental or ecological assessments without a good deal of biochemical and physiological knowledge.

Studies of the effects of toxic metals in the environment also benefit from parallel studies in the relatively controlled conditions of the laboratory, because, no matter how careful we are, effects seen in the wild could be due, not to the metal, but to some factor of which we were unaware. Laboratory work can do much to help confirm that the effects seen in the wild really were due to the chemical or metal concerned.

From the preceding paragraphs, it is apparent that a multi-disciplinary approach is required to understand the toxic effects of metals. So, I thought it would be valuable to bring together a number of research workers who, because they belong to groups working on such diverse aspects of metal problems, might not ever have an opportunity of meeting.

In designing the workshop, 3 areas of work seemed particularly important and interesting:-

1. Studies on the accumulation of metals by wildlife and on the effects of the metals.

2. Studies attempting to determine the localisation of metals in cells.

3. Studies on the metallothioneins, those ubiquitous but intriguing proteins whose role in trace element metabolism and metal toxicology still defies definition more than 20 years after the protein was first described.

Only the reader will be able to judge whether the combination of expertise represented in the papers in this volume helps to show how inter-laboratory co-operation and a multi-disciplinary approach to metals can increase our knowledge and understanding. If nothing else, the papers show what is being done across a broad range of specialities, and, in that context, it is of interest that 4 of the 5 UK research councils (ARC, MRC, NERC and SERC) have supported the efforts of the workers who took part in this meeting.

D Osborn
November 1983

1. ACCUMULATION AND EFFECTS OF METALS

FOOD CHAIN RELATIONSHIPS OF COPPER AND CADMIUM IN HERBIVOROUS AND INSECTIVOROUS SMALL MAMMALS

B A HUNTER*, M S JOHNSON** and D J THOMPSON*
*Department of Zoology
**Department of Botany
University of Liverpool, P O Box 147, Liverpool L69 3BX

ABSTRACT

This paper contrasts the transfer of copper and cadmium through the food chains of contaminated grasslands. Particular emphasis is placed on the accumulation of metals by endemic populations of herbivorous and insectivorous small mammals. Copper was found to be highly mobile through the soil-plant-invertebrate system. However, natural homeostatic mechanisms prevent significant copper accumulation in small mammals. In contrast, cadmium shows much greater mobility and bioconcentration throughout the food chain. By virtue of respective dietary contamination levels, the insectivorous common shrew accumulates 10-fold more cadmium than the herbivorous field vole occupying the same contaminated grasslands. The important influences of diet, trophic status and metal speciation on food chain transfer and relative accumulation of metals by small mammals are discussed.

INTRODUCTION

Trace metal accumulation in small mammals endemic to metal contaminated environments has been reported by several authors (Roberts & Johnson 1978). The present paper examines copper and cadmium movement through the grassland food web. Transfer of metals through the soil-plant-invertebrate pathway to small mammals is described, with particular emphasis on the influence of diet on metal bioavailability and accumulation by small mammals. For this reason, the study concentrates on 2 species of small mammal which exhibit contrasting dietary habits and trophic status. These are the herbivorous field vole (*Microtus agrestis* L.) and the insectivorous common shrew (*Sorex araneus* L.). Both species are widespread and common in British grasslands and have been proposed as bioindicators for monitoring food chain metal contamination (Beardsley *et al.* 1978). Consequently, although the present data describe grasslands in the vicinity of a copper refinery, the results could be applied, with care, to other environments such as those contaminated by the mining and refining of metalliferous ores where soils may show a similar range of metal contamination levels.

MATERIALS AND METHODS

Sample collection was carried out in contaminated, semi-contaminated and control environments. Contaminated grasslands were in close proximity to a major copper refinery housing a copper-cadmium alloying plant. The semi-contaminated site was 1 km away from the refinery in the direction of the marginally prevailing wind, and the control site was located well away from both urban and industrial sources of metal contamination.

A series of 0-5 cm surface soil cores from each site provided a basic index of site contamination levels. Endemic vegetation and ground surface dwelling invertebrates were collected throughout the year at monthly intervals at each site. Small mammal trapping was carried out using baited live-traps. In the laboratory, all samples were subjected to wet oxidation using concentrated nitric acid at 120^0C and the resulting digests were analysed for copper and cadmium using standard atomic absorption techniques.

RESULTS AND DISCUSSION

Metal concentrations in surface soils from the control site were consistent with reported data for unpolluted soils elsewhere (Thornton 1980). In comparison with control site values, surface soils from the refinery and 1 km sites both showed highly significant (p<0.001) elevation of copper and cadmium concentrations (Table 1). The ratio of Cu : Cd in contaminated surface soils which represent the base of food chain accumulation was approximately 1000 : 1. Transect samples have shown an exponential fall-off in soil metal concentrations away from this refinery (Hunter & Johnson 1982).

At the refinery site, elevated copper concentrations in surface soils have inhibited microbial degradation of organic matter, resulting in a pronounced build-up of undecomposed plant litter. In addition, the high copper concentrations have placed a marked selection pressure on the endemic vegetation. Consequently, plant diversity has been reduced to copper tolerant populations of just 3 species. Of these, 2 are grasses - creeping bentgrass (*Agrostis stolonifera* L.) and red fescue (*Festuca rubra* L.), and the third is field horsetail (*Equisetum arvense* L.). Dietary analysis has shown all 3 species to be important food plants for *M. agrestis* in the refinery grasslands.

Vegetation metal concentrations increase in line with site and soil contamination levels. All 3 plant species showed significantly elevated copper and cadmium concentrations at the refinery and 1 km sites (Table 2). In the refinery grasslands, there are 2 major sources of metal contamination in plants: first, root uptake of metals and subsequent translocation to the shoots and, second, deposition of metal particulates from the atmosphere and their adherence to exposed leaf surfaces. Experiments have shown that over 70% of the metals measured in unwashed refinery site vegetation are present as particulates attached to leaf surfaces. The extent of the particulate contamination varies according to leaf surface morphology, the period of leaf surface exposure and the distance from the refinery. Leaf surface morphology influences particulate retention, which follows the order *E. arvense* > *A. stolonifera* > *F. rubra*, and contributes to interspecific differences in metal concentrations. Timing of new growth and the exposure period produces a marked seasonal cycle of contamination levels, which shows a pronounced winter peak in vegetation copper and cadmium concentrations. Examination of leaf surface particulates using X-ray microanalysis has revealed a chemical composition typical of resuspended, contaminated soil surface dusts. The diet of *M. agrestis* consists of the fresh green stems and leaves of grasses and horsetails (Ferns 1976). Clearly, contaminated site *M. agrestis* will ingest protein bound/ionic metals from plant

TABLE 1   Concentrations of copper and cadmium in surface soils
from contaminated and control sites

	Copper	Cadmium
Refinery	11025.0 ± 1592.0***	15.5 ± 2.3***
1 km site	543.0 ± 51.5***	6.9 ± 0.6***
Control	15.1 ± 0.9	0.8 ± 0.1

Data in mg/kg ± SE (dry wt); n = 40 samples in all cases.
*** denotes significant difference at p < 0.001 between control
and contaminated soils.

TABLE 2   Annual mean copper and cadmium concentrations in *Agrostis
stolonifera*, *Festuca rubra* and *Equisetum arvense* from
contaminated and control sites

	Copper		
	A. stol.	*F. rub.*	*E. arv.*
Refinery	121.9 ± 31.6***	73.3 ± 12.4***	138.9 ± 21.7***
1 km site	24.5 ± 2.4***	22.9 ± 2.5***	28.9 ± 1.8***
Control	9.9 ± 0.7	7.5 ± 0.6	14.3 ± 0.9
	Cadmium		
	A. stol.	*F. rub.*	*E. arv.*
Refinery	3.3 ± 0.36***	2.9 ± 0.3 ***	2.6 ± 0.24***
1 km site	1.3 ± 0.07***	1.3 ± 0.09***	1.0 ± 0.08***
Control	0.6 ± 0.06	0.5 ± 0.05	0.5 ± 0.04

Data in mg/kg ± SE (dry wt); n < 200 in all cases.
*** denotes significant difference at p < 0.001 between control
and contaminated sites.

TABLE 3   Annual mean copper and cadmium concentrations in beetles
and spiders from contaminated and control sites

	Copper		Cadmium	
	Beetles	Spiders	Beetles	Spiders
Refinery	466.0 ± 108.0***	952.0 ± 148.0***	14.6 ± 1.56***	95.2 ± 8.6***
1 km site	52.0 ± 3.0***	180.0 ± 14.0***	5.2 ± 0.7 ***	26.0 ± 3.15***
Control	25.0 ± 1.3	73.0 ± 7.4	0.6 ± 0.1	2.5 ± 0.3

Data in mg/kg ± SE (dry wt); n < 300 in all cases.
*** denotes significant difference at p < 0.001 between control
and contaminated sites.

tissue, together with metal-silicate and oxide particulates attached to leaf surfaces. These metal species will exhibit widely differing absorption coefficients in the mammalian gut, suggesting that metal concentrations alone, without metal speciation data, are inadequate criteria on which to assess the likely effects of dietary exposure.

In contrast, the diet of *S. araneus* covers a wide range of litter and ground surface dwelling invertebrates (Pernetta 1976). However, the diet is predominantly beetles and spiders. For this reason, beetle and spider data alone have been used to represent the diet of *S. araneus*. The Coleoptera data cover the Carabidae and Staphylinidae while those for Araneida include the Linyphiidae and Lycosidae.

Concentrations of copper and cadmium are significantly elevated in beetles and spiders at both contaminated sites (Table 3). This effect illustrates the mobility of both metals in the invertebrate food web. Cadmium shows further marked accumulation in invertebrates, particularly spiders. By its position in the food chain, the ratio of Cu : Cd has been reduced from 1000 : 1 in the contaminated substrate to 10 : 1 in spiders. The result indicates a remarkable 100-fold difference in the food chain mobility of copper and cadmium from contaminated soil through to predatory invertebrate species.

The relationships between dietary copper and cadmium concentrations and total body metal concentration for *M. agrestis* and *S. araneus* at contaminated and control sites are illustrated in Figure 1. By feeding on invertebrates, *S. araneus* ingests 3x more copper and 12x more cadmium than *M. agrestis* feeding on grasses in the same contaminated grasslands, calculated on a mg/kg/d basis. In *M. agrestis*, total body copper concentrations remain constant despite a 10-fold increase in dietary copper intake (mg/kg/d). *S. araneus*, however, shows an approximate doubling in total body copper concentration at the refinery site in response to an increase in dietary intake rate from 16 to 260 mg/kg/d between control and refinery sites respectively. Such stability reflects the remarkable efficiency of the mechanisms for Cu-homeostasis in mammals (Evans 1973). In contrast, *Microtus* and *Sorex* both show significant cadmium accumulation in relation to their respective diets. This accumulation is particularly evident in *S. araneus* which shows a marked increase of this non-essential trace metal. The importance of metal speciation in respect of absorption and accumulation of metals in mammals has already been mentioned. The accumulation of cadmium by *S. araneus* from a highly contaminated invertebrate diet is compounded by a pronounced change in the bioavailability of cadmium in ingested invertebrates. It would appear that cadmium ingested in a protein bound form from invertebrates is more freely absorbed from the alimentary canal than cadmium ingested as plant material.

Once absorbed, cadmium in small mammals is bound to metallothionein and over 80% of the total body burden is located in the liver and kidney target organs. Contaminated populations of *S. araneus* and *M. agrestis* show significant age-accumulation of Cd-thionein in the liver and kidney (Hunter *et al.* 1981). Toxicological symptoms of cadmium accumulation in *S. araneus* are described in a subsequent paper in this volume.

9

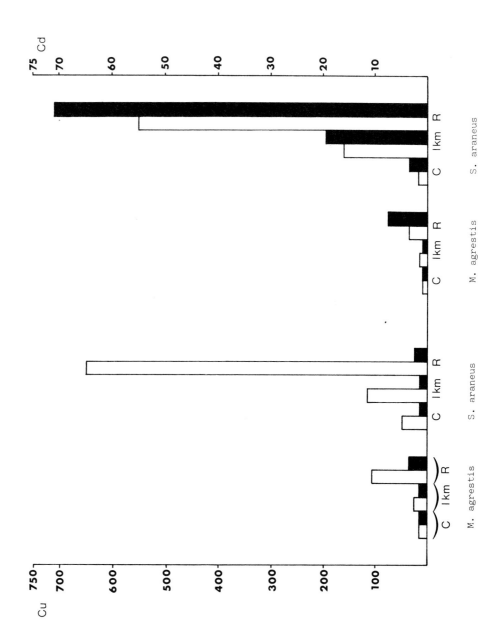

Figure 1 Relationship between copper and cadmium concentrations in M. agrestis and
S. araneus (mg/kg dry wt.) and their respective herbivorous and insectivorous
diets. Solid columns: total body concentration; empty columns: dietary
concentration. C = control site, 1 km = 1 km site, R = refinery site.

CONCLUSION

Analysis of copper and cadmium in soils, vegetation, invertebrates and small mammals from contaminated and control environments has revealed marked differences in the potential for food chain transfer. Copper is mobile in the soil-plant-invertebrate system and shows particular accumulation within the invertebrate food web. Further bioconcentration of copper in insectivorous common shrews is buffered by highly efficient homeostatic control mechanisms.

Cadmium, however, shows markedly different environmental behaviour. The food chain mobility of cadmium from soil to invertebrate predator is 100 times greater than that of copper. Cadmium accumulates at each step in the food chain, including small mammals. Through dietary habits, the insectivorous common shrew ingests much greater amounts of cadmium than the herbivorous field vole in the contaminated refinery grasslands. Consequently, shrews show marked cadmium accumulation in the liver and kidney target organs.

REFERENCES

BEARDSLEY, A., VAGG, M.J., BECKETT, P.H.T. & SANSOM, B.F. 1978. Use of the field vole (*M. agrestis*) for monitoring potentially harmful elements in the environment. *Environ. Pollut.*, 16, 65-71.

EVANS, C.W. 1973. Copper homeostasis in the mammalian system. *Physiol. Rev.*, 53, 535-570.

FERNS, P.N. 1976. Diet of a *Microtus agrestis* population in S.W. Britain. *Oikos*, 27, 506-511.

HUNTER, B.A. & JOHNSON, M.S. 1982. Food chain relationships of copper and cadmium in contaminated grassland ecosystems. *Oikos*, 38, 108-117.

HUNTER, B.A., JOHNSON, M.S., THOMPSON, D.J. & HOLDEN, H. 1981. Age accumulation of copper and cadmium in wild populations of small mammals. In: *Proc. int. Conf. Heavy Metals in the Environment Amsterdam, 1981*, 263-266. Edinburgh: CEP Consultants.

PERNETTA, J.C. 1976. Diet of the shrews *Sorex araneus* and *Sorex minutus* in Wytham grassland. *J. Anim. Ecol.*, 45, 899-912.

ROBERTS, R.D. & JOHNSON, M.S. 1978. Dispersal of heavy metals from abandoned mine workings and their transference through terrestrial food chains. *Environ. Pollut.*, 16, 293-310.

THORNTON, I. 1980. Background levels of heavy metals in soils and plants. In: *Reclamation of contaminated land*. Paper C5. London: Society of Chemical Industry.

CADMIUM WITHIN A CONTAMINATED GRASSLAND ECOSYSTEM ESTABLISHED ON METALLIFEROUS MINE WASTE

S M ANDREWS and J A COOKE
Biology Department, Sunderland Polytechnic, Sunderland SR1 3SD

ABSTRACT

Cadmium was taken up by vegetation and accumulated in higher trophic levels of the invertebrate food web of a grassland ecosystem established on metalliferous mine waste. Total body concentrations of cadmium in the herbivorous small mammal *M. agrestis* L. and the insectivore *S. araneus* L. were related to the cadmium concentration in their estimated diets, being significantly higher than those found in the same small mammal species at a control site. The principal target organ for cadmium accumulation was the kidney in *M. agrestis* at both sites and *S. araneus* at the control site, but the liver adopted this role with the higher body concentrations in *S. araneus* from the mine site.

INTRODUCTION

The establishment and maintenance of grassland are the major goals of many metalliferous mine waste reclamation schemes. Often the intention is to create a semi-natural ecosystem in order to encourage further recolonization by wildlife. However, it is important to assess the consequences for plant and animal communities of long term exposure to mining residues. Even modern operations produce waste with relatively high concentrations of metallic minerals that can undergo weathering reactions which may increase their environmental mobility. Cadmium is one such element that can be highly toxic and is capable of bioaccumulation through terrestrial food webs (Martin & Coughtrey 1975, 1976). This paper describes the distribution of cadmium within a contaminated grassland ecosystem established on metalliferous mine waste at a mining complex in central England. Revegetation was carried out several years ago and the waste, although containing high residual concentrations of lead, zinc and cadmium, supports a dense grass sward. It has a complex invertebrate food web, a diverse avian fauna and also supports breeding populations of the short-tailed field vole (*Microtus agrestis* L.), and the common shrew (*Sorex araneus* L.). The control site was an area of similar rough grassland distant from any roads or industrial sources of cadmium.

Samples of composite vegetation and surface soil were collected from both sites. Roots were separated out from soil blocks by hand and washed with deionized water to remove adhering soil particles. Plant litter was also collected. Ground living macroinvertebrates were sampled using pitfall traps and small mammals were caught using baited Longworth traps.

Soil and vegetation were dried and finely ground (< 0.5 mm) invertebrates were sorted into major taxonomic groups and oven dried. Small mammals were dissected to remove liver and kidneys which together with the remaining carcase were freeze-dried. All tissues were digested in concentrated nitric acid (AnalaR): soils at 120^0C for 4 hours; vegetation and animal tissues at 100^0C for one hour. Cadmium determinations were carried out using atomic absorption spectrophotometry.

RESULTS AND DISCUSSION

Cadmium concentrations of soil and vegetation at the control site were within the accepted range for unpolluted sites (Peterson & Alloway 1979) and were significantly lower than the values for the mine site (Table 1).

TABLE 1   Concentration of Cd (mg/kg dry wt, mean ± SE) in surface soil and vegetation. No. of samples in parentheses

	Control site	Mine site
Surface soil (top 5 cm)	0.52 ± 0.03 (12)	23.90 ± 1.79** (8)
Live grass leaf	1.21 ± 0.14 (23)	4.74 ± 1.26** (28)
Grass flowering stem	1.20 ± 0.36 (11)	7.89 ± 2.79** (13)
Root	3.62 ± 0.72 (6)	18.88 ± 1.82** (6)
Litter	2.04 ± 0.74 (6)	8.97 ± 1.24** (6)

** denotes significant difference between sites ($p < 0.001$).

The concentrations of cadmium in the vegetation at the mine site were lower than the surface soil and decreased in the order root > stem > leaves. This finding is similar to another recent study (Matthews & Thornton 1982).

Invertebrate cadmium concentrations were significantly elevated at the mine site (Table 2) and were generally greater than the concentration in vegetation or soil.

TABLE 2   Concentration of Cd (mg/kg dry wt, mean ± SE) in selected invertebrate groups. No. of samples in parentheses

	Control site	Mine site
Carabidae	0.50 ± 0.08 (17)	6.75 ± 0.69** (22)
Araneae and Opiliones	2.44 ± 0.37 (15)	29.13 ± 1.54** (10)
Diptera	2.88 ± 0.53 (14)	33.04 ± 3.48** (11)
Lumbricus spp	7.06 ± 1.67 (11)	67.62 ± 3.09** (11)

** denotes   significant difference between sites ($p < 0.001$).

Table 3 represents the estimated dietary cadmium concentrations for the 2 species of small mammal, the herbivore *Microtus agrestis* and the insectivore *Sorex araneus* at both sites (for discussion of the calculation of estimated diets see Hunter & Johnson 1982 and references cited therein). At the mine site, the dietary concentrations of cadmium were higher for both species but with that of *Sorex araneus* being about 5-fold higher than *Microtus agrestis*.

The diet of *S. araneus* consists to a large extent of species representing higher trophic levels of the invertebrate food web, through which cadmium can accumulate. Thus, the shrew's rapacious appetite for a diet which happens to be high in cadmium leads to a higher total body concentration, at the mine site, 28-fold that of *M. agrestis* feeding less voraciously on a diet (green vegetation) of lower cadmium concentration (Table 3). Although cadmium body burdens are largely derived from dietary intake, grooming, respiration and foetal burdens are additional contributory sources.

TABLE 3   Concentrations of Cd (mg/kg dry wt, mean ± SE) in *M. agrestis* and *S. araneus* and their estimated diets. No. of samples in parentheses

		Estimated diet	Total body concentration
Control site	*M. agrestis*	1.21	0.88 ± 0.05 (20)
	S. araneus	2.1	1.19 ± 0.07 (13)++
Mine site	*M. agrestis*	4.7	1.84 ± 0.12 (21)**
	S. araneus	23.2	52.7 ± 3.4 (17)++ **

** Significant difference between sites ($p < 0.001$).
++ Significant difference between species at the same site ($p < 0.001$).

Kidney and liver tissues are known to be the main sites of cadmium accumulation in vertebrates. Therefore, these were chosen for further study and results are shown in Table 4. The kidney and liver concentrations for both species of small mammal were significantly elevated at the mine site. The characteristic pattern of tissue concentrations, with kidney being higher than the liver, was demonstrated by both species at the control site and by *M. agrestis* at the mine site. However, in *S. araneus* at the mine site the liver concentration was much higher than that of the kidneys.

If a comparison is made of the total organ burden ratios (kidney/liver), then at the control site for both species the kidneys contain about 50% of the cadmium burden found in the liver (Table 4). In *M. agrestis* at the mine site, feeding on a diet of slightly elevated cadmium concentration, the cadmium burden in the kidneys increased to approximately the same as in the liver. However, in *S. araneus* at the mine site, with a relatively high cadmium diet, the liver retained a much higher proportion of cadmium compared to the kidney and had higher concentrations than the kidneys (Table 4).

TABLE 4   Kidney and liver Cd concentrations (mg/kg, mean ± SE)
and kidney : liver total organ burden (µg Cd) ratios in
M. agrestis and *S. araneus*. No. of samples in parentheses

| | *M. agrestis* | | *S. araneus* | |
	Control site (20)	Mine site (21)	Control site (13)	Mine site (17)
Kidney	1.75 ± 0.26	5.21 ± 0.86**	4.13 ± 0.72	158 ± 15.4**
Liver	1.14 ± 0.15	1.88 ± 0.22*	2.85 ± 0.55	235.9 ± 29.7**
Kidney : liver ratio	0.45	0.97	0.5	0.2

* and ** denote significant differences between sites at $p < 0.01$ and
$p < 0.001$ respectively.

Figure 1 shows how the percentage of the total body burden contained
in the kidneys and liver varies with the total body burden for *S. araneus*
at the mine site. The kidneys show a consistent percentage of the total
body burden (approx. 5%) even though the body burden increases from 66 µg
cadmium to 236 µg cadmium. In contrast, the liver's percentage increased
from 18% to 35% within the same range of total body burdens, which shows
clearly the importance of the liver as a target organ with increasing body
burdens of cadmium in *S. araneus*.

Figure 1   Regression of % of total body burden in liver and kidney
against total body burden in *S. araneus* at the mine site

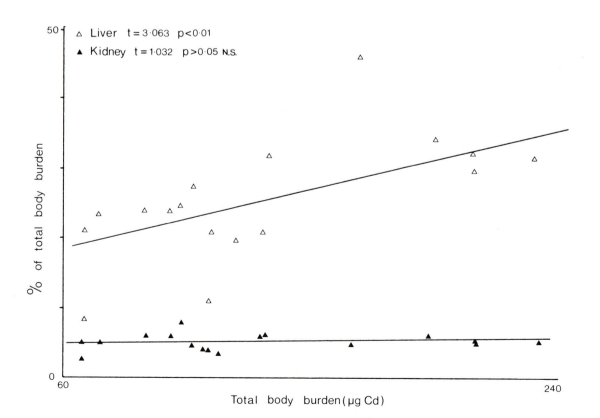

CONCLUSIONS

This study has demonstrated that residual cadmium in the substrate can be taken up by vegetation and accumulated in higher trophic levels of a grassland ecosystem established on metalliferous mine tailings. Total body concentrations of both *M. agrestis* and *S. araneus* were related to their estimated dietary cadmium concentrations. Especially high concentrations of cadmium were found in liver and kidneys of *S. araneus* at the mine site. These concentrations in *S. araenus* are higher than some that have produced cadmium toxicity in laboratory experiments (Decker *et al*. 1958). Similar symptoms, therefore, may well exist within this wild shrew population and the relative change in target organ status from kidney to liver may be a consequence of kidney damage as this organ is known to be particularly sensitive to cadmium.

REFERENCES

DECKER, L.E., BYERRUM, R.U., DECKER, C.F., HOPPERT, C.A. & LANGHAM, R.F. 1958. Cadmium toxicity studies I. Cadmium administered in drinking water to rats. *Arch. Ind. Health,* 18, 228-231.

HUNTER, B.A. & JOHNSON, M.S. 1982. Food chain relationships of copper and cadmium in contaminated grassland ecosystems. *Oikos,* 38, 108-117.

MARTIN, M.H. & COUGHTREY, P.J. 1975. Preliminary observations on the levels of cadmium in a contaminated environment. *Chemosphere,* 4, 155-160.

MARTIN, M.H. & COUGHTREY, P.J. 1976. Comparisons between the levels of lead, zinc and cadmium within a contaminated environment. *Chemosphere,* 5, 15-20.

MATTHEWS, H. & THORNTON, I. 1982. Cadmium in the farming environment in Shipham. In: *Proc. int. Cadmium Conf., 3rd, Miami, 1981,* edited by D. Wilson & R.A. Volpe, 91-82. London: Cadmium Association; New York: Cadmium Council and International Lead and Zinc Research Organization.

PETERSON, P.J. & ALLOWAY, B.J. 1979. Cadmium in soils and vegetation. In: *The chemistry, biochemistry and biology of cadmium,* edited by M. Webb, 45-92. Amsterdam: Elsevier/North Holland.

LEAD POISONING IN THE MUTE SWAN

M BIRKHEAD
EGI, Department of Zoology, South Parks Road, Oxford OX1 3PS

ABSTRACT

In England, lead poisoning through the ingestion of anglers' weights is the major cause of mortality in the mute swan. On the River Thames, 143 swans have been examined at post-mortem and 58% died from lead poisoning. Kidney proved to be the most useful indicator tissue. Examination of the lead levels of live birds on the River Thames showed very few swans (< 50%) had levels below the maximum acceptable level of 40 µg/100 ml. Blood lead levels of flock birds were lowest during the close coarse fishing season and highest during the summer months. Cygnet mortality was significantly higher on the lower Thames where blood lead levels were known to be at their highest. Ultrastructural examinations revealed electron dense renal intranuclear inclusions and abnormalities in the liver and erythrocytes. X-ray microprobe analysis demonstrated that the electron dense granules in the kidney were lead.

INTRODUCTION

In Great Britain, the mute swan (*Cygnus olor*) is a royal bird; the Crown can lay claim to any swan unless it possesses a unique mark of ownership. The Crown also has the right to say who may or may not own a swan and the question of ownership has been taken seriously for centuries (Birkhead & Perrins 1981). Today, it is only the swans on the River Thames which are still caught and marked during the annual ritual known as 'swan-upping'. In 1979, 2 studies were established to examine the decline in numbers of the mute swan. The first study was initiated by the then Minister for the Environment who noticed the gradual disappearance of a local flock at Stratford-upon-Avon, and, as a result, a nationwide enquiry was launched to determine factors involved in the decline in numbers (Nature Conservancy Council 1981). At the same time, the owners of the Thames swans, the Crown and the worshipful companies of the Vintners and Dyers, sponsored the present study, which examined the decline in numbers and causes of mortality of the mute swan on the River Thames (Birkhead & Perrins 1981; Birkhead 1982a). It was clear from both enquiries that the major cause of mortality was lead poisoning due to ingestion of anglers' discarded lead weights (NCC 1981; Birkhead 1981, 1982a) and that certain populations in England were declining as a result (Hardman & Cooper 1980; Birkhead & Perrins 1981). Lead poisoning is also known to be a major mortality factor in other species of wildfowl but the source of lead is usually lost gun-shot (Roscoe 1978; Thomas 1980).

This paper aims to bring together 3 separate aspects concerning lead poisoning and swans which have been dealt with more fully elsewhere:

1.   Causes of mortality (Birkhead 1982a)
2.   Lead levels in live swans (Birkhead 1983)
3.   Ultrastructural examinations (Birkhead 1982b)

RESULTS

1.   <u>Causes of mortality</u>. Between September 1979 and October 1982, 143 mute swans were examined at post-mortem. Lead poisoning through the ingestion of anglers' lead weights accounted for 58% of the deaths. In addition, 2% of the birds died as a result of lost fishing line, floats or hooks. Kidney proved to be the most useful indicator tissue; the median lead level in birds with lead weights in their gizzard was 908 mg/kg dry wt compared to 8 in birds without weights. Swans dying from lead poisoning were characteristically emaciated: mean ($\pm$ 1 SE) weight of lead poisoned birds 5.7 $\pm$ 0.23 kg compared to 8.9 $\pm$ 0.39 kg ($p \ll 0.001$) in birds dying from other causes.

   There was a clear distinction in the seasonal distribution of recorded death between swans dying from lead poisoning and those dying from other factors (Table 1). More birds died from lead poisoning during the months of July-October, whilst other causes of mortality, such as collisions with power cables, took their highest toll during the winter months of November-February (Table 1).

2.   <u>Lead levels in live swans</u>. One hundred mute swans were bled from Abbotsbury, Dorset, where there is no fishing. All those birds had blood lead levels below 40 µg/100 ml and this was adopted as the maximum acceptable blood lead level, MAL.

   Lead levels of flock birds (non-breeding immatures aged between 1-4) increased with proximity to London. Birds on the tributaries and upper Thames had median blood lead levels below or around 40 µg/100 ml, whilst swans in Windsor and Richmond had blood lead levels of around 100 µg/100 ml (Figure 1).

   Regular bleeding of 10 swans a month at Reading-on-Thames showed significant seasonal variation, with the lowest levels being recorded during the close fishing season and the highest during the summer months (Figure 2).

   Breeding females had significantly higher blood lead levels than mated males on the same stretch of river, and birds on the River Thames had significantly higher lead levels than those on the tributaries (Table 2).

   Cygnets from the River Thames had higher blood lead levels than those from the tributaries (Table 3). There was no trend with proximity to London but brood sizes were significantly higher on the tributaries where blood lead levels were at their lowest.

3.   <u>Ultrastructural examinations</u>. Two swans were collected for ultra-structural examination (Table 4). Renal intra-nuclear inclusions were present in the lead poisoned swan and X-ray microprobe analysis

THE MARYLAND LIBRARY
RD., LIVERPOOL, L18 8JD

demonstrated that these granules consisted of an amorphous mass of
lead (Figure 3). Large numbers of electron dense granules were
observed in the liver of the lead poisoned bird and occasionally
in the healthy bird's liver (Figure 4). X-ray analysis demonstrated
these granules contained iron, probably as part of haemosiderin.
The presence of these iron granules in the liver is probably due
to lead's interference in the ferrochetalase system, preventing the
incorporation of iron into the haemoglobin molecule. In such cases,
zinc takes the place of iron in some porphyrin molecules and, as a
result, iron may be incorporated into the granules in the liver.
Electron dense granules were also observed in erythrocytes of the
lead poisoned swan, lying in the cytoplasm just inside the plasma
membrane. These were not observed in the healthy swan's blood.
Their composition was not determined, although it seems likely they
were either lead or iron.

DISCUSSION

Lead poisoning through the ingestion of anglers' lead weights was
the major cause of mortality of the mute swan on the River Thames. The
choice of tissue used to confirm lead poisoning has been discussed by
Alder (1944) and Simpson et al. (1979), and in this study kidney proved
to be the most useful indicator tissue. It has been suggested that a
kidney lead level of at least 125 mg/kg dry wt is required to confirm
plumbism (Clarke & Clarke 1975) and the results of this study would
support such a claim.

Flock bird blood lead levels increased with proximity to London, a
phenomenon also shown in pigeons (Columba livia) (Hutton 1980). Hutton
(1980) suggests this was due to an increase in ingesting food contaminated
with roadside dust.   It is unlikely to be the explanation in mute swans
and the increase in background lead levels (that is below 40 µg/100 ml)
with proximity to London was probably caused by a number of factors. This
gradual increase in blood lead levels is illustrated by the breeding males
with levels ranging from 20 µg on the tributaries to 40 µg/100 ml on the
lower Thames. The much greater increase in blood lead levels of flock
birds may be associated with the availability of lost lead weights. Most
birds caught on the lower Thames are caught in urban flocks, virtually
the only way swans exist in these areas. The urban swans may be
particularly prone to lead poisoning for a number of reasons. One is
that weights lost in these areas remain available on the concrete wharfs
and embankments, whereas weights lost in the countryside are more likely
to disappear in the mud and grassy banks (Simpson et al. 1979; Birkhead
1981). Boating may also add to the problem by stirring up the sediment
which contains lead weights, and, as the river is tidal at places like
Richmond-on-Thames, weights lost on the bank may get washed into the
shallows following high tide.

Of the incubating birds, females from the tributaries had the lowest
levels, but no regional trend could be detected on the River Thames. At
this time of year, females had significantly higher lead levels than males,
and this could be explained in a number of different ways. Prior to egg
formation, females require calcium in large quantities to form eggs. They
can obtain this either by mobilizing stored calcium from bones (Taylor &

TABLE 1   Monthly distribution of deaths for November 1979–
         October 1981

	November– February	March– June	July– October
Lead poisoned	11	9	34
Other than lead poisoned	21	7	8

$\chi^2 = 16.5$, 2 df, $p < 0.001$

The monthly breakdown has been devised to include the period March–
June which includes the coarse fishing 'close season'. A more
complete data set is presented in Figure 1.

TABLE 2

(a) Median blood lead levels (μg/100 ml) of incubating females

Region	1979	1980	1981	Range	n
Abbotsbury	30			18–81	
Tributaries		39[a]	33[a]	4–820	30
Upper Thames		120[b]	92[b]	25–436	19
Lower Thames		113[b]	59[b]	8–271	25

(b) Median blood lead levels (μg/100 ml) of males when females
    were incubating

Region	1979	1980	1981	Range	n
Abbotsbury	29			19–57	
Tributaries		14[a]	20[a]	5–37	17
Upper Thames		31[a,c]	31[a,c]	2–64	12
Lower Thames		40[c]	42[c]	6–65	8

All statistical comparisons presented in these Tables were confined
to within-sex differences in the same year. No superscript in common
denotes statistical difference at the 5% level (Mann-Whitney U-test).

Males and females from Abbotsbury did not have significantly different
lead levels from one another (U = 91, Mann-Whitney U-test). Females
from all other areas had significantly higher lead levels than males
from the same areas ($p < 0.05$ Mann-Whitney U-test).

TABLE 3 Cygnet lead levels and mean brood sizes at the end of the season

Lead levels (μg/100 ml)	1979 Abbotsbury	1980 Thames	1980 Tributaries	1981 Thames	1981 Tributaries
0-20	80	20	61	3	51
21-40	7	12	9	9	16
41-60	–	10	3	11	3
61-80	–	11	3	7	0
81-100	–	3	0	6	0
101-1000	–	9	4	29	2
1000+	–	1	1	–	0
Median	8	37	10	92	12
Range	1-33	0-1684	1-3730	15-458	0.6-145
Mean brood size	–	2.66	4.40	2.39	3.67

TABLE 4 Details of the 2 swans taken for ultra-structure examination

Sex	Weight (kg)	Blood lead μg/g 100 ml	Liver lead mg/kg dry wt	Kidney lead mg/kg dry wt	No. shot in gizzard
Male	8.7	57	4	8	0
Male	6.0	354	120	1085	5

Figure 1 Flock bird lead levels (μg/100 ml) on the River Thames. Region 1, upper Thames and tributaries; 1980 n = 16 (range 8-48); 1981 n = 18 (range 5-67). Region 2, the Thames from Oxford to Goring; 1980 n = 29 (range 10-269); 1981 n = 18 (range 17-2870). Region 3, the Thames from Goring to Windsor; 1980 n = 39 (range 9-458); 1981 n = 28 (range 50-3866). Region 4, the Thames from Windsor to Richmond; 1980 n = 24 (range 16-1482); 1981 n = 14 (range 77-213)

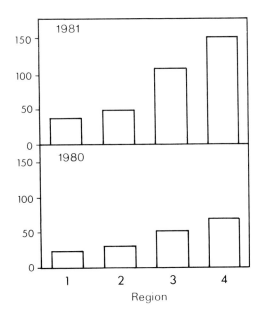

Figure 2a Monthly median blood lead levels from Reading-on-Thames, 1980, showed significant heterogeneity; H = 32.2, p < 0.001, Kruskal-Wallis test

Figure 2b Monthly median blood lead levels from Reading-on-Thames, 1981, showed significant heterogeneity; H = 52.8, p < 0.001, Kruskal-Wallis test

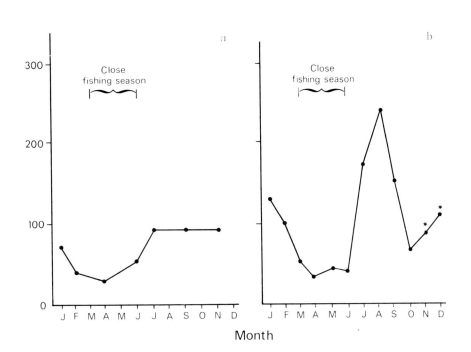

Figure 3   Section of kidney of the lead poisoned swan
                showing the electron dense intra-nuclear
                inclusion. x19 500

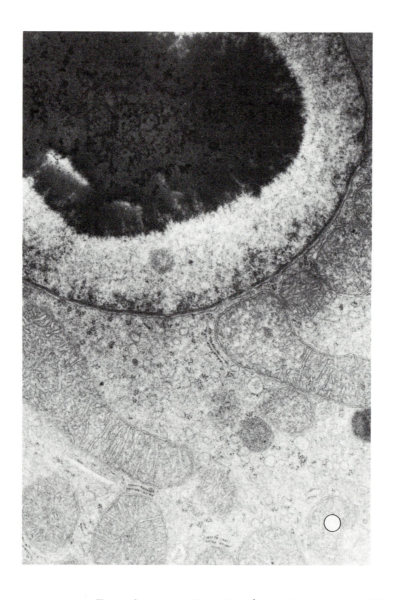

Figure 4   Two photographs showing the extraordinary dense
                nature of the granules (arrowed) associated
                with the hepatocytes of the lead poisoned swan.
                x21 000

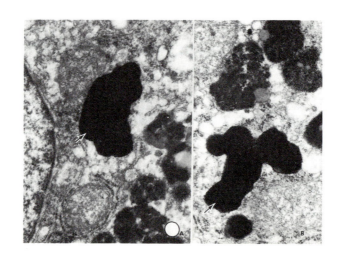

Moore 1954; Taylor 1970) or by absorbing more from the alimentary canal (Baltrop & Khoo 1976). Lead is also stored in bone and follows similar metabolic pathways to calcium (Clarke & Clarke 1975); consequently, females may exhibit elevated blood lead levels at this time of year if (i) they mobilize calcium, and therefore lead, from bone; or (ii) they selectively absorb more calcium, and as a result absorb more lead from the alimentary canal.

The ultrastructural examinations revealed the usual intra-nuclear inclusions associated with lead poisoning (Locke *et al.* 1966; Simpson *et al.* 1979; Hutton 1980). In addition, X-ray analysis confirmed the locality of the inclusions and that they consisted of lead. The inclusions appeared to be confined to the nucleolus as was the case in pigeons (*Columba livia*) (Hutton 1980). However, no mitochondrial disorders were observed (Hutton 1980). The granules in the liver were confirmed as iron, a phenomenon also noted in whistling swans (*Cygnus columbianus*) affected by lead (Trainer & Hunt 1965).

REFERENCES

ALDER, F. 1944. Chemical analyses of organs from lead poisoned Canada geese. *J. Wildl. Mgmt*, **8**, 83-85.

BALTROP, D. & KHOO, H.E. 1976. The influence of dietary minerals and fat absorption of lead. *Sci. Total Environ.*, **6**, 265-273.

BIRKHEAD, M.E. 1981. How the fishermen kill the swans. *New Scient.*, **90**, 75-77.

BIRKHEAD, M.E. 1982a. Causes of mortality in the mute swan on the River Thames. *J. Zool.*, **198**, 15-25.

BIRKHEAD, M.E. 1982b. Intracellular localization of lead in tissues of the mute swan. *Tissue & Cell*, **14**, 691-701.

BIRKHEAD, M.E. 1983. Lead levels in the blood of mute swans on the River Thames. *J. Zool.*, **199**, 59-73.

BIRKHEAD, M.E. & PERRINS, C.M. 1981. The decline of the royal swan. *New Scient.*, **91**, 75-77.

CLARKE, E.G. & CLARKE, M.L. 1975. *Veterinary toxicology*. 3rd ed. London: Baillere.

HARDMAN, J. & COOPER, D.R. 1980. Mute swans on the Warwickshire Avon - a study of a decline. *Wildfowl*, **31**, 29-36.

HUTTON, M. 1980. Metal contamination of feral pigeons from the London area: Part 2 - biological effects of lead exposure. *Environ. Pollut. A*, **22**, 281-293.

LOCKE, L.N., BAGLEY, G.E. & IRBY, H.D. 1966. Acidfast intranuclear bodies in the kidneys of mallards fed lead shot. *Bull. Wildl. Dis. Ass.*, **2**, 127-131.

NATURE CONSERVANCY COUNCIL. 1981. *Lead poisoning in swans*. London: NCC.

ROSCOE, D.E. 1978. *Pathology and plumbism in waterfowl and development of a simple diagnostic blood test*. Unpubl. PhD thesis, University of Connecticut.

SIMPSON, V.R., HUNT, A.E. & FRENCH, M.C. 1979. Chronic lead poisoning in a herd of mute swans. *Environ. Pollut.*, **18**, 187-202.

TAYLOR, T.G. 1970. The role of skeleton in egg shell formation. *Ann. Biol. anim. Biochem. Biophys.*, **10**, 83-91.

TAYLOR, T.G. & MOORE, J.H. 1954. Skeleton depletion in hens laying on a low calcium diet. *Br. J. Nutr.*, 8, 112-124.

THOMAS, G.J. 1980. A review of ingested lead poisoning in wildfowl. *Bull. int. Waterfowl Res. Bur.*, 46, 43-60.

TRAINER, D.O. & HUNT, R.A. 1965. Lead poisoning of whistling swans in Wisconsin. *Avian Dis.*, 9, 252-264.

LEAD POISONING IN MUTE SWANS - AN EAST ANGLIAN SURVEY

M C FRENCH
Institute of Terrestrial Ecology, Monks Wood Experimental Station, Abbots Ripton, Huntingdon, Cambs PE17 2LS

ABSTRACT

From October 1981 to September 1982, 320 dead mute swans (*Cygnus olor*) were examined and total lead levels were determined in liver, kidney, and bone. Chemical analyses revealed that 70% (224) of these swans had died of lead poisoning. In the majority of cases, death was due to an acute exposure. Thirty-six dead birds and 8 birds which were given a veterinary examination before they died had tissue levels of lead which did not conform to the accepted levels that would confirm fatal lead poisoning. These birds were included in the total of lead poisoned birds after taking 3 further points into account: (i) veterinary diagnosis; (ii) post-mortem evidence, including the presence of lead fishing weights in the gizzard; and (iii) the effect of a fall in body temperature usually experienced by lead poisoned birds prior to death which was made worse by the very low environmental temperatures prevailing at the time these 44 birds were examined.

INTRODUCTION

The poisoning of wildfowl by the ingestion of spent gun-shot pellets is well documented throughout Europe and North America (Belrose 1959; Del Bono 1970; Clausen *et al*. 1975; Beer & Stanley 1964; French 1982). In Britain, a joint study carried out in 1973 by ITE and the Veterinary Investigation Service in Nottingham was able to identify for the first time that ingested anglers' weights were responsible for the deaths of mute swans on the River Trent in Nottingham (Simpson *et al*. 1979). In America, several common loons (*Garia immer*) died after ingesting lead anglers' weights (Locke *et al*. 1982). Swans and other waterfowl are assumed to take in the lead weights along with, or in mistake for, grit which they need to aid digestion. Following the joint ITE-VI Centre study, the NCC report on lead poisoning in swans was published (NCC 1981). This report identified lead poisoning from anglers' weights as being the biggest single cause of mute swan deaths in Great Britain. It also identified several 'hot spots' throughout Britain where 70-90% of reported swan deaths were due to lead poisoning. In some areas, previously substantial swan populations have declined markedly, eg River Thames (Birkhead 1982), and in others they have disappeared completely (eg River Avon near Stratford).

This report outlines some results of a project investigating the effects of lead on swans in East Anglia. A major objective of this study is to determine the proportion of swans dying through the ingestion of lost or discarded fishing weights.

METHODS

Between October 1981 and September 1982, a study was done on the rivers Welland, Nene, Ouse, and Cam, a total river length of approximately 300 miles. Also included were gravel pits and reservoirs within the same area. During this time, 320 mute swan (*Cygnus olor*) carcases were examined.

Chemical analyses of liver, kidney, and bone were done to help establish the cause of death. The contents of the gizzard and proventriculus were examined and pieces of lead shot identified by eye. In the majority of cases where lead shot was found, it was possible to distinguish visually between fishing weights and spent gun-shot. Where there was doubt, the material was tested chemically to differentiate between the 2 sources. Only 4 mute swans were shown to have carried gun-shot in their gizzards. These 4 birds were eliminated from the study, even though 2 contained anglers' weights as well as the spent gun-shot.

RESULTS AND DISCUSSION

The majority of the swans were corpses with no known history, but in 8 cases clinical observations were available following X-ray and veterinary examination. The symptoms in these 8 cases were anorexia, weakness, inability to move correctly and impaction of the oesophagus. These birds were very emaciated. These are all signs of lead poisoning; in addition, a marked and sustained fall in body temperature, observed up to 72 hours prior to death, was noted in the 8 live birds. A similar drop in body temperature has been demonstrated with pigeons dosed with lead fishing weights in this laboratory. The body weights of all 320 birds in the study are given in Table 1. These findings are consistent with our original observations (Simpson *et al.* 1979).

TABLE 1   Mute swan body weights (kg)

	Lead poisoned (lead shot present)	Lead poisoned (lead shot absent)	Cause of death other than lead poisoning
Number	174	50	96
Range	(5.1-9.8)	(4.2-7.9)	(8.1-13.2)
Mean	7.1	6.0	9.8

At post-mortem, sexes were determined and 70% (224) of the birds examined were female. The sexes and monthly receipts of swan bodies are shown in Figure 1.

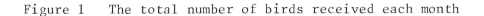

Figure 1   The total number of birds received each month

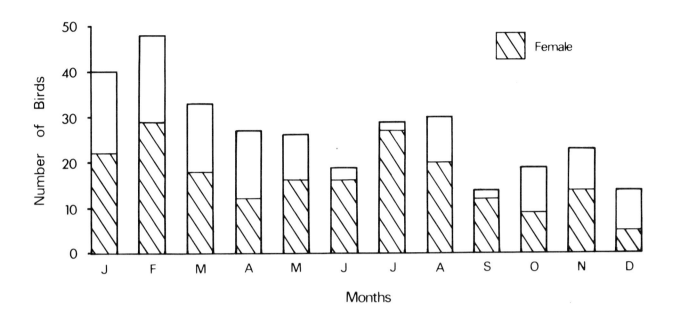

Lead shot was present in the alimentary tract of 174 swans and in all cases death was due to lead poisoning. Fishing hooks and attached line were found in 6 birds and in only one case could these have contributed to death. None of these 6 birds died of lead poisoning.

Chemical analyses of kidney, liver, and bone showed that 70% (224) of the birds examined had died of lead poisoning. Of the 224, 180 birds in the survey had levels of lead in the tissues in excess of the values that would confirm fatal lead poisoning in mammals. These values are 125 mg/kg dry wt lead in the kidney and 50 mg/kg dry wt lead in the liver (Clarke & Clarke 1975). Forty-four birds with lower levels of lead in their tissues were also included in the lead poisoned group, because the post-mortem and veterinary evidence obtained were classically those of lead poisoned birds. The levels of lead in these 44 swans were (expressed as mg/kg dry wt): 13-32 (mean 28) in liver, 38-64 (mean 53) in kidney, and 40-86 (mean 63) in bone. However, at post-mortem these birds exhibited the accepted signs of lead poisoning, ie impaction of the oesophagus, proventriculus and gizzard, wasting of the muscle, a high level of iron in the liver, an enlarged gall bladder and the presence of lead fishing weights in the gizzard. In addition, 8 of these 44 swans were the live birds that received a veterinary examination prior to death, and whose classical signs of lead intoxication have already been mentioned.

One explanation as to why these swans died with relatively low levels of lead in their tissues could be that they died in the winter of 1981-82 when temperatures of -17^0C were recorded in the East Anglian region. These cold conditions could have exacerbated the hypothermia that may be associated with lead poisoning in birds, making it impossible for the birds to recover or causing them to succumb more easily.

The deposition of lead in the intoxicated and dead birds can provide evidence of the type of exposure that has taken place, namely a high bone and low soft tissue level indicates a chronic exposure and a low bone, high soft tissue level indicates an acute form of exposure (Clarke & Clarke 1967). In all but 20 birds, kidney lead levels were higher than bone levels, indicating that death was due, in a majority of cases, to acute exposure.

The high number of swans reported dead and confirmed as lead poisoned in this study supports previous findings from other areas (NCC 1981; Birkhead 1982).

The number of female and young non-breeding swans dying of lead poisoning is difficult to explain. Lead retention in rats is known to be associated with age (Shields *et al*. 1978). Young rats fed lead acetate retained more lead at a given dietary intake than mature rats. This work may go someway to explain our observations. Breeding females need calcium for egg production and this can be furnished in several ways by mobilising stored calcium in the skeleton - up to 40% of the stored calcium can be used in this way (Sturkie 1954) - or the need can be met from an increased absorption from the alimentary canal (Barltrop & Khoo 1976). Because the metabolism of lead follows closely that of calcium, a high demand for calcium yields a high availability of lead to the soft tissues. Female ducks in breeding condition are known to accumulate more lead in their tissues than do males (Finley *et al*. 1976; Finley & Dieter 1978).

Finally, we could find no influence of the close fishing season on the availability of lead to swans. We received more lead poisoned birds during the close season than in the equivalent previous or following period of time. Swans can take up to 3 weeks to die from lead intoxication after ingesting fishing weights, so this fact must be taken into account when interpreting the close season data.

REFERENCES

BARLTROP, D. & KHOO, H.E. 1976. The influence of dietary minerals and fat on the absorption of lead. *Sci. Total Environ.*, 6, 265-273.
BEER, J.V. & STANLEY, P. 1964. Lead poisoning in the Slimbridge Wildfowl collection. *Rep. Wildfowl Trust, 16th, 1963-64*, 30-34.
BELROSE, F.C. 1959. Lead poisoning as a mortality factor in waterfowl populations. *Bull. Ill. St. nat. Hist. Surv.*, 27, 235-288.
BIRKHEAD, M.E. 1982. Causes of mortality in the mute swan on the River Thames. *J. Zool.*, 198, 15-25.
CLARKE, E.G.C. & CLARKE, M.L. 1967. *Garner's veterinary toxicology*. 3rd ed. New York: Williams & Wilkins.
CLARKE, E.G.C. & CLARKE, M.L. 1975. *Veterinary toxicology*. 3rd ed. London: Baillière.
CLAUSEN, A.G., DALSGAARD, H. & WOLSTRUP, C. 1975. Udbrid af blyforgiftning blandt danske knopsvaner (*Cygnus olor*). *Dansk Vet. Tidsskr.*, 21, 843-847.
DEL BONO, G. 1970. Il saturnismo degli uccelli acquatici. *Annali Fac. Med. vet. Univ. Pisa*, 23, 102-151.

FINLEY, M.T. & DIETER, M.P. 1978. Influence of laying on lead accumulation in bone of mallard ducks. *J. Toxicol. & environ. Health,* *4,* 123-129.

FINLEY, M.T. DIETER, M.P. & LOCKE, L.N. 1976. Lead in tissues of mallard ducks dosed with two types of lead shot. *Bull. environ. Contam. & Toxicol.,* *16,* 261-269.

FRENCH, M.C. 1982. Lead poisoning in Bewick swans. *BTO News,* no.121, 1.

LOCKE, L.N., KERR, S.M. & ZOROMSKI, D. 1982. Lead poisoning in common loons (*Gavia immer*). *Avian Dis.,* *26,* 392-396.

NATURE CONSERVANCY COUNCIL 1981. *Lead poisoning in swans.* London: NCC.

SHEILDS, J.B., MITCHELL, H.H. & RUTH, W.A. 1978. The metabolism and retention of lead in growing and adult rats. *J. Ind. Hyg. Toxicol.,* *21*(1), 7-23.

SIMPSON, V.R., HUNT, A.E. & FRENCH, M.C. 1979. Chronic lead poisoning in a herd of mute swans. *Environ. Pollut.,* *18,* 187-202.

STURKIE, P.D. 1954. *Avian physiology.* New York: Comstock Publishing Associates.

CADMIUM AND MERCURY IN SEABIRDS

D OSBORN* and J K NICHOLSON**

*Institute of Terrestrial Ecology, Monks Wood Experimental Station, Abbots Ripton, Huntingdon, Cambs PE17 2LS

**Department of Anatomy, St Thomas Hospital Medical School, London SE1 7EH[†]

ABSTRACT

At some colonies, seabirds can contain high levels of cadmium and mercury. Tissue metal levels are higher at some periods of the year than at others. Very high levels can be found in individuals which are members of breeding pairs and which could, therefore, be considered as 'apparently healthy'. Biochemical studies on seabirds with high toxic metal levels have shown that, in common with other vertebrates, they contain a metallothionein-like protein in their liver and kidney. However, despite the presence of this 'protective' protein, ultrastructural studies revealed that the kidneys of these seabirds had various types of nephrotoxic lesions which could be reproduced in the laboratory by dosing birds with cadmium and mercury. These findings open a number of questions about the significance of the toxic metal levels found in some colonies of seabirds.

INTRODUCTION

When this study began in 1976, there were relatively few data on the levels of metals in seabirds and no information on their possible significance. However, it was clear from unpublished work done earlier at Monks Wood that, on occasion, seabirds found dead on the shore could contain far higher levels of cadmium than were found in other species of birds. This observation raised the question as to what significance toxic metals had for seabirds, especially if the metals were accumulated as a result of pollution.

To investigate this question, it was necessary (i) to determine what the metal levels were in seabirds from a colony far from sources of pollution (as a natural 'control' observation); (ii) to determine whether seabirds contained metallothionein when they contained high metal levels; and (iii) to try to determine whether high levels of toxic metals were in any way deleterious to the birds. This report summarises the results of our investigations on seabirds to date.

[†]Present address: Toxicology Unit, School of Pharmacy, Brunswick Square, London WC1

METHODS

Much of the work has concerned 3 species of birds, collected from 2 colonies. The colonies were St Kilda - off the west coast of Scotland in the north-east Atlantic, and the Isle of May - off the east coast of Scotland on the edge of the North Sea. The species studied were puffin (*Fratercula arctica*), fulmar (*Fulmarus glacialis*) and Manx shearwater (*Puffinus puffinus*).

Other procedures are described in papers referred to elsewhere in this report.

RESULTS

The initial data from St Kilda, largely obtained from birds that were members of breeding pairs, were surprising in that higher levels of cadmium were found in tissues of these birds than had then been reported in any other wild vertebrate (Bull *et al.* 1977). Substantial amounts of mercury were also found (Murton *et al.* 1978). These findings were confirmed in a subsequent study which also showed that the cadmium was not confined to the liver and kidney but was found in other organs, such as gonad, pancreas and intestine. In addition, the mercury was present in the form of methyl mercury (Osborn *et al.* 1979), generally regarded as more toxic than inorganic mercury - partly because it may not be as readily excreted.

A metallothionein-like protein was found in the tissues of the fulmar which, while binding much of the cadmium in the tissue, bound very little of the mercury (Osborn 1978). Although there was a body of evidence to suggest that metallothionein might protect animals from the effects of metals, it seemed worthwhile checking to see if kidney damage was associated with the high metal levels. To make such a study fully effective, it was necessary to examine not only birds contaminated with high levels of metals, but also birds that contained relatively little metal. Osborn (1979) described the intercolony differences in tissue metal concentrations that existed between puffins from St Kilda and those from the Isle of May. Although the Isle of May is in the North Sea (often regarded as more polluted than the Atlantic), puffins from this colony had much lower levels of both cadmium and mercury than did puffins from St Kilda (Figure 1). Further, a seasonal trend in metal concentrations was observed in the Isle of May puffins, with the highest metal concentrations occurring at the beginning of the breeding season (Osborn 1979). This suggested that there would be times when metal levels in St Kilda seabirds were even higher than those recorded, as the St Kilda birds were collected only in the middle of their breeding period.

This regional difference meant seabirds from both a 'high' and a 'low' metal colony could be obtained, and a comparison made of their tissue metal levels and the histological appearance of their kidneys. In addition, birds were dosed in the laboratory with cadmium and mercury, to guard against the possibility that any unusual features seen in seabird kidneys were caused by some factor other than cadmium or mercury.

32

Figure 1 Metal concentrations in puffin tissues. More details in Osborn (1979), Osborn *et al.* (1979)

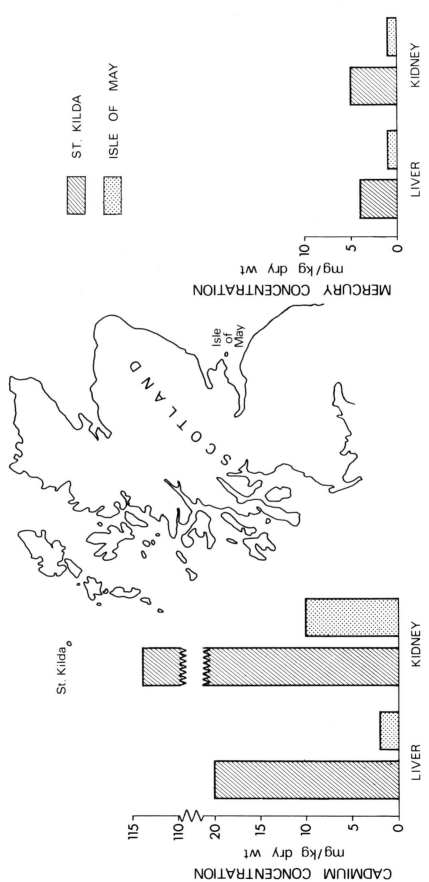

Accounts of the observations on the seabird and metal dosed bird kidneys have now been published (Nicholson & Osborn 1983) and the results are summarised in Table 1. In brief, St Kilda birds exhibited numerous nephrotoxic lesions, while no abnormalities were observed in birds from the 'low' metal site on the Isle of May. Metal induced lesions in laboratory birds were strikingly similar to those of the St Kilda birds and occurred at similar tissue levels of metals in both groups of animals.

DISCUSSION

To some extent, these results leave our original question about the significance of high metal levels in seabirds unanswered. We now know that high metal levels are found at various seabird sites around the globe (Anderlini *et al.* 1972; Bull *et al.* 1977; Osborn *et al.* 1979; Furness & Hutton 1979; Stoneburner & Harrison 1981) and it is becoming an accepted view – that still requires more investigation – that the metal arises from natural rather than anthropogenic sources.

This study has shown that metallothionein-like proteins cannot prevent tissue damage occurring, although they may reduce its extent allowing tissue repair mechanisms time to come into operation. Indeed, evidence for tissue regeneration was obtained in both seabirds and metal dosed birds (Nicholson & Osborn 1983). These results raise questions about the role of metallothionein in metal toxicity.

Further research is needed to determine the significance of the high metal levels for seabirds, and a number of factors not mentioned here must be taken into account, including (i) the influence of age and sex on metal accumulation; and (ii) the role of nutritional status on metal binding in the tissues (Bremner & Davies 1975) – a potentially important variable in animals like seabirds which may feed little while incubating eggs for days at a time. Also, this report has taken no account of the levels of essential metals such as zinc, copper and iron, although some of these have been subjects of study (eg Osborn *et al.* 1979; Osborn 1979). To understand the effects of toxic metals on seabirds there can be little doubt that a sound knowledge of essential metal metabolism in birds must be obtained.

In short, there is much to be done before we can say whether high metal levels have effects on seabirds which are ecologically significant, ie before we can answer the question: do metals influence either the breeding success or survival prospects of individuals and/or different populations of seabirds?

REFERENCES

ANDERLINI, V.C., CONNORS, P.G., RISEBOROUGH, R.W. & MARTIN, J.H. 1972. Concentrations of heavy metals in some Atlantic and North American seabirds. In: *Proc. Colloq. on Conservation Problems in Antarctica*, 49-62. Blacksburg: Virginia Polytechnic Institute & State University.

BREMNER, I. & DAVIES, N.T. 1975. The induction of metallothionein in rat liver by zinc injection and restriction of food intake. *Biochem. J.*, <u>149</u>, 733-738.

BULL, K.R., MURTON, R.K., OSBORN, D., WARD, P. & CHENG, L. 1977. High levels of cadmium in Atlantic seabirds and seaskaters. *Nature, Lond.*, <u>269</u>, 507-509.

FURNESS, R. & HUTTON, M. 1979. Pollutant levels in the great skua *Catharacta skua*. *Environ. Pollut.*, <u>19</u>, 261-268.

MURTON, R.K., OSBORN, D. & WARD, P. 1978. Are heavy metals pollutants in Atlantic seabirds? *Ibis*, <u>120</u>, 106-107.

NICHOLSON, J.K. & OSBORN, D. 1983. Kidney lesions in pelagic seabirds with high tissue levels of cadmium and mercury. *J. Zool.*, <u>200</u>, 99-118.

OSBORN, D. 1978. A naturally occurring cadmium and zinc binding protein from the liver and kidney of *Fulmarus glacialis* a pelagic north Atlantic seabird. *Biochem. Pharmac.*, <u>27</u>, 822-824.

OSBORN, D. 1979. The significance of metal residues in wild animals. In: *Proc. int. Conf. Management and Control of Heavy Metals in the Environment, 1979*, 187-192. Edinburgh: CEP Consultants.

OSBORN, D., HARRIS, M.P. & NICHOLSON, J.K. 1979. The comparative tissue distribution of mercury, cadmium and zinc in three species of pelagic seabirds. *Comp. Biochem. Physiol. C*, <u>64</u>, 61-67.

STONEBURNER, D.L. & HARRISON, C.S. 1981. Heavy metal residues in sooty tern tissues from the gulf of Mexico and north central Pacific Ocean. *Sci. Total Environ.*, <u>17</u>, 51-58.

TABLE 1   Pathological features observed in St Kilda seabirds

Severity and types of pathological features observed	Puffin	Manx shearwater	Fulmar
Renal corpuscles			
Necrosis of Bowman's capsule cells	−	+	+
Podocyte vacuolation and nuclear crenation	+	++	++
Mesangial matrix, changes and abnormal electron lucency	−	+	+
Proximal tubules			
Dilation of extracellular spaces	+++	+	+
†Mitochondrial swelling (intramatrical)	+	++	++
Nuclear pyknosis	−	++	++
Degree of cell necrosis	+	++	++
Obstruction of distal tubules, etc, by necrotic debris	+	++	++

†Varied considerably from cell to cell.

Severity of changes and degree of damage: − no detectable abnormalities; + few and/or slight changes; ++ common and/or moderately severe changes; +++ very frequent and/or severe changes.

MERSEY ESTUARY BIRD MORTALITIES

D OSBORN, K R BULL and WENDY J YOUNG
Institute of Terrestrial Ecology, Monks Wood Experimental Station, Abbots Ripton, Huntingdon, Cambs PE17 2LS

ABSTRACT

Between late summer and early winter in the years 1979-1982, birds died on the Mersey estuary with high levels of alkyl lead compounds in their tissues. The behavioural and morphological changes seen in these polluted birds in the wild were also seen in birds dosed with alkyl lead compounds in the laboratory. Furthermore, tissue levels of alkyl lead compounds in affected laboratory birds were similar to those found in affected wild birds. We concluded that the majority of birds that were found dead on the Mersey in this period were killed by the alkyl lead. A monitoring programme to determine the levels of alkyl lead in birds now living on the estuary has shown that many birds contained sufficient alkyl lead to cause sub-lethal effects and to impair their chances of survival. Further mortalities may occur should environmental factors, linked with the hydrodynamics of the estuary area, combine to cause an increase in the birds' exposure to alkyl lead compounds.

INTRODUCTION

The Mersey estuary in north-west England is heavily industrialised and receives effluent from these industries and sewage water from the towns and cities in the area. The estuary supports a large number of birds, mainly overwintering waders, wildfowl and gulls. At least 2500 birds were found dead on the estuary in 1979, and smaller mortalities occurred in 1980, 1981 and 1982. High levels of alkyl lead compounds were found in the birds. Earlier reports have described the incidents between 1979 and 1981 (Head *et al.* 1980; Osborn & Bull 1982; Bull *et al.* 1983). Details of experimental studies examining the toxicity of alkyl lead compounds to birds have also been published (Osborn *et al.* 1983). These experimental studies supported the view that the majority of the birds found dead on the estuary in recent years were killed by alkyl lead compounds.

This report outlines the main findings of the incident-related work on birds, gives data on the levels of alkyl lead in birds shot and netted on the estuary for monitoring purposes, and briefly discusses possible and actual effects of these levels on the birds.

METHODS

Methods have been described in Bull *et al.* (1983) and Osborn *et al.* (1983).

THE MARKLAND LIBRARY
STAND PARK RD., LIVERPOOL L16 8JD

RESULTS

As described by Bull *et al.* (1983), mortalities on the Mersey estuary involving large or notable numbers of birds have occurred between late summer and early winter.

Many species have been involved. In 1979, 1300 of the birds found dead were dunlin (*Calidris alpina*) along with several hundred gulls (mainly black-headed gulls (*Larus ridibundus*)). In 1980, mortalities were mainly gulls, *L. ridibundus* again. In 1981, mortalities were spread amongst several species and in 1982 herons (*Ardea cinerea*) were prominent. There have been wildfowl casualties in all years (eg teal (*Anas crecca*) and mallard (*Anas platyrhynchos*)).

Affected birds on the estuary exhibited unco-ordinated movements and a head tremor. Some seemed unable to feed properly. Post-mortem examinations of dead birds showed they had discoloured intestines, brilliant green bile and discoloured livers. Analyses for toxic chemicals showed that the only measurable chemical detected in signifi-cant quantities was an alkyl lead compound. This was most probably trimethyl lead. Tissue levels of alkyl lead were generally >10 mg/kg wet wt in the livers of dead birds. In shot and netted birds, some contained >5 mg/kg wet wt and a high proportion contained >1 mg/kg. Many of the shot and netted birds had some of the abnormal morphological features seen in dead birds. Osborn *et al.* (1983) showed that birds that died when dosed with alkyl lead compounds exhibited very similar behavioural and morphological features to those seen in sick and dying birds on the Mersey. Birds given sub-lethal doses of alkyl lead compounds showed loss of condition and had the morphological features found in live birds shot and netted on the Mersey. In addition, the experimental work showed that dosing with alkyl lead compounds led to dose-related enlargement of the gall bladder and dose-related bone marrow activation.

Tissue levels of alkyl lead in dosed birds were similar to those observed in wild birds in which similar effects had been seen.

It was concluded that the experimental work supported the view that alkyl lead compounds had poisoned birds on the Mersey estuary and that the majority of the bird deaths recorded on the Mersey between 1979 and 1982 were caused by the birds eating prey contaminated with these compounds.

Attempts have been made to collect birds from the estuary at intervals so that (i) the amounts of alkyl lead in their tissues could be measured and (ii) they could be examined to see if they showed any of the abnormal morphological features experimentally associated with sub-lethal tissue levels of alkyl lead. Attention has been concentrated on 2 species, teal and dunlin.

Figure 1 shows alkyl lead levels in Mersey teal livers between 1980 and 1982. Although there may be some evidence of a decline in levels (we believe alkyl lead effluent levels on the estuary have fallen), some birds still contain high enough levels to cause some sub-lethal effect. Seventeen teal collected in February and November 1982 have been studied in particular detail to look not only for morphological abnormalities, but also for evidence of the loss of condition that alkyl lead compounds brought about in laboratory birds. No evidence of any loss of condition was found in these teal, but the majority exhibited some morphological abnormality in that they had enlarged gall bladders and green-stained livers and intestines. Teal affected in this way had a mean level of alkyl lead in their liver greater than that found in the unaffected birds (Figure 1).

For various reasons, fewer dunlin samples have been obtained. Figure 1 also shows the liver levels of alkyl lead found in dunlin netted on the Mersey. Overall, these show little real sign of any decline in the level in live birds since the time of the 1979 incident, although when the August 1980 results became available it appeared a decline had occurred. Furthermore, morphological changes in dunlin were more severe than those in teal. Studies of dunlin body condition have yet to be made.

DISCUSSION

There are a number of features of these mortality incidents that can only properly be considered when studies currently in hand at the North West Water Authority laboratories, and elsewhere, are completed. These features include an explanation of why mortalities were not seen in earlier years and why they tend to occur in the period of the year when they do. They may all be explained in terms of the hydrodynamics of the estuary and associated waterways.

However, so far as the birds are concerned, many still have levels of alkyl lead in their tissues that are sufficient to cause morphological changes. Also, mean levels in teal and dunlin are not yet below the 0.5 mg/kg wet wt which had been suggested (Bull *et al.* 1983; Osborn *et al.* 1983) as the maximum level above which levels should not rise if the chance of a substantial further mortality was to be avoided with a fair degree of certainty.

It seems that large numbers of birds on the estuary (particularly dunlin and other waders) are still at risk from alkyl lead compounds and that, given an unfortunate combination of environmental factors linked with the hydrodynamics of the estuary area, another substantial bird mortality could occur.

ACKNOWLEDGEMENTS

We thank Mr R Cockbain for dunlin samples and Mr D Jones and his colleagues of the Frodsham and District Wildfowlers (BASC) for the teal. This work was supported by Nature Conservancy Council funds as part of the programme of research into nature conservation.

REFERENCES

BULL, K.R., EVERY, W.J., FREESTONE, P., HALL, J.R., OSBORN, D., COOKE, A.S. & STOWE, T.J.   1983.   Alkyl lead pollution and bird mortalities on the Mersey estuary, U.K., 1979-1981. *Environ. Pollut. A*, <u>31</u>, 239-259.

HEAD, P.C., d'ARCY, B.J. & OSBALDESTON, P.J.   1980.   *The Mersey estuary bird mortality autumn-winter 1979 - Preliminary report*.   North West Water Authority.   Directorate of Scientific Services.   Scientific Report Ref. no. DSS-EST-80-1.

OSBORN, D. & BULL, K.R.   1982.   Mersey bird mortalities 1979-1981: a pollution problem resolved. *Annu. Rep. Inst. terr. Ecol. 1981*, 28-33.

OSBORN, D., EVERY, W.J. & BULL, K.R.   1983.   The toxicity of trialkyl lead compounds to birds. *Environ. Pollut. A*, <u>31</u>, 261-275.

Figure 1   Concentrations of alkyl lead in livers of live-caught teal (left) and dunlin (right)

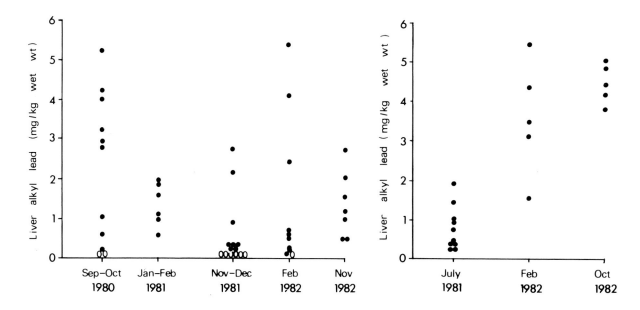

Notes:

1. More birds must be analysed before statistical tests can be done or conclusions about trends in the data can be drawn.

2. 0 = <0.1 mg/kg.

3. Many of these birds showed internal signs of being affected by alkyl lead (Osborn *et al.* 1983).   In 17 teal examined in February and November 1982, alkyl lead levels in livers of the 10 affected animals were 1.7 mg/kg compared to 1.0 mg/kg for unaffected birds.   The respective geometric means were 1.1 and 0.4 mg/kg.

4. Some data were added after the workshop.

II. LOCALISATION OF METALS IN ANIMAL TISSUES

CADMIUM INDUCED LESIONS IN TISSUES CF *SOREX ARANEUS* FROM METAL REFINERY GRASSLANDS

B A HUNTER*, M S JOHNSON** and D J THOMPSON*
*Department of Zoology
**Department of Botany
University of Liverpool, P O Box 147, Liverpool L69 3BX

ABSTRACT

Wild common shrews endemic to metal contaminated grasslands contain some of the highest concentrations of metals recorded in wildlife. Accumulated copper and cadmium are centred in the liver and kidneys. Electron microscope examination has revealed widespread liver and kidney damage. Lesions in the kidney affect glomerular and proximal tubule cells, while in the liver hepatocytes and sinusoids show most damage. Hepatocytes contained cytoplasmic cadmium inclusion bodies. Sinusoidal lumens showed large scale inflammatory cell invasion, and the presence of immature red blood cells indicated a form of cadmium induced anaemia.

INTRODUCTION

Food chain relationships of copper and cadmium in metal contaminated and control grasslands are described in a preceding paper in this volume. Feeding habits and high metabolic rate combine to amplify cadmium exposure in the insectivorous common shrew (*Sorex araneus* L.) (Hunter & Johnson 1982). Consequently, tissue cadmium concentrations in *S. araneus* are amongst the highest that have been recorded in wildlife. Cadmium accumulation is age-related and is centred on the liver and kidney target organs, where over 93% of the total body burden is stored (Hunter *et al.* 1981). Cadmium is an element with no ascribed biological role and is toxic to most enzyme systems (Underwood 1977). Any toxicological symptoms of accumulated cadmium would therefore be expected to develop in the liver and kidney target organs. This paper describes cadmium induced lesions in the liver and kidney of *S. araneus* endemic to contaminated environments. The results are considered in relation to existing data from laboratory investigations.

MATERIAL AND METHODS

Common shrews were live-trapped using blow fly pupae as bait; traps were checked at 6 hour intervals. Specimens were transferred to perspex cages and fed on blow fly larvae and assorted invertebrates from the sites of capture. Contaminated animals were captured from grasslands within a major copper refinery housing a copper-cadmium alloying plant; control animals were caught at a site distant from both urban and industrial sources of metal contamination. Liver and kidney tissues were subjected to immersion or perfusion fixation for 3 hours in 2% paraformaldehyde and 2.5% gluteraldehyde in 0.1 M cacodylate buffer. This treatment was followed by

post-fixation in 1% osmium tetroxide. Tissues were ultramicrotome
sectioned to 60-90 nm and post-stained with uranyl acetate and lead
citrate. Examination was carried out on an AEI Corinth 500 transmission
electron microscope. Portions of each tissue were analysed for copper
and cadmium using standard atomic absorption techniques.

RESULTS AND DISCUSSION

Both copper and cadmium concentrations were significantly elevated
in *S. araneus* from the refinery site. Liver and kidney copper
concentrations in refinery populations were approximately double those
recorded in control animals ($p < 0.05$ and < 0.01 respectively) (Table 1).
Copper accumulation may occur following failure of the normal homeostatic mechanisms due to the high dietary exposure at the refinery site,
or due to disrupted copper metabolism resulting from the accumulated
cadmium burden (Evans *et al.* 1970). Cadmium shows marked accumulation
($p < 0.001$) in the liver and kidney target organs (Table 2). Tissue
cadmium concentrations ranging from 300-1000 mg/kg dry wt in the liver
and 150-560 mg/kg in the kidney are amongst the highest recorded in
terrestrial wildlife. These concentrations represent storage of 87.0%
and 5.7% of the total body cadmium burden in the liver and kidney
respectively.

TABLE 1   Whole body and tissue copper concentrations in
Sorex araneus from refinery and control grasslands

	Whole body	Liver	Kidney
Refinery	22.8 ± 2.5**	63.1 ± 14.4*	51.0 ± 5.7**
Control	12.1 ± 0.9	23.7 ± 1.9	30.7 ± 1.8

Data in mg/kg ± SE dry wt.
* and ** denote statistically significant differences at $p < 0.05$
and < 0.01 respectively between refinery and control population.

TABLE 2   Whole body and tissue cadmium concentrations in
Sorex araneus from refinery and control grasslands

	Whole body	Liver	Kidney
Refinery	47.5 ± 10.5***	577.4 ± 124.0***	252.8 ± 75.1***
Control	3.6 ± 0.3	13.6 ± 1.4	20.5 ± 1.6

Data mg/kg ± SE dry wt.
*** denotes statistically significant differences at $p < 0.001$ between
refinery and control populations.

Electron microscopy has revealed widespread tissue damage in the liver and kidneys. In the kidneys, lesions were evident in both glomerular and proximal tubule cells. Glomerular damage (Figure 1) was manifest as thickening and proliferation of basement membrane material, disruption and fusion of epithelial podocytic processes and hypertrophy of endothelial and mesangial cells. The presence of fibrin was noted in several glomeruli. However, despite the presence of ultrastructural damage in some glomeruli, others showed normal morphology throughout the range of cadmium intoxication. The frequency of proximal tubule cell damage increased in parallel with tissue cadmium concentration. In particular, the frequency of electron dense cells increased. Such cells were not recorded in control specimens.

Toxicological lesions in proximal tubule cells included increased cytoplasmic density, swollen and distorted mitochondria and nuclei with invaginations and increased karyoplasmic density (Figure 2). The basement membrane was thickened and there was prolific basal invagination of cell membranes. A characteristic feature of cadmium damaged proximal tubule cells was a marked proliferation of enlarged apical cytoplasmic vesicles and numerous autophagic lysosomes. The microvillus brush border and lumen had a normal appearance. There was no evidence of regeneration in proximal tubule tissues.

In contrast with some previous studies, kidney damage was remarkably slight. More severe damage has been recorded in wild sea-birds and experimentally dosed starlings, despite much lower kidney cadmium concentrations than those found in *S. araneus* (Nicholson & Osborn 1983). This may suggest a degree of adaptation to high cadmium levels in *S. araneus* endemic to these contaminated grasslands.

Nevertheless, liver tissue showed even more complex and widespread damage, the severity of which was correlated with cadmium concentration. Hepatocytes from juvenile *S. araneus* showed disrupted rough endoplasmic reticulum (RER), abnormal and swollen mitochondria, dilation of the smooth endoplasmic reticulum (SER) and invagination of nuclei. Cells contained electron dense cytoplasmic inclusion bodies only infrequently. In adults, hepatocytes were less disrupted and the RER, SER and mitochondria appeared normal.

Hepatocytes from adults contained numerous electron dense cyto-plasmic inclusion bodies, shown by X-ray microanalysis to consist of cadmium, sulphur and selenium (Figure 3). Further liver lesions were apparent in the sinusoids. Endothelial cells showed marked hypertrophy and degeneration of the space of Disse. Sinusoidal lumens were swollen and packed with inflammatory and plasma cells including macrophages, neutrophils, thrombocytes and immature red blood cells such as reticulocytes and normoblasts (Figure 4). The presence of immature red blood cells indicates a form of cadmium induced anaemia, but there was no evidence of haemopoietic activity. It is unclear whether the release of plasmacytes, resembling plasma cell hepatitis, results from a toxic effect of cadmium on the bone marrow or a direct effect on mature erythrocytes (Page & Good 1960). Cadmium induced liver lesions of this magnitude have not previously been recorded. However, it would appear that cadmium inclusion bodies ameliorate hepatocyte toxicity in *S. araneus*. Tissue cadmium concentrations increase with age yet

42

Figure 2   Proximal tubule damage in cadmium
           contaminated kidney from *S. araneus*
           x5000

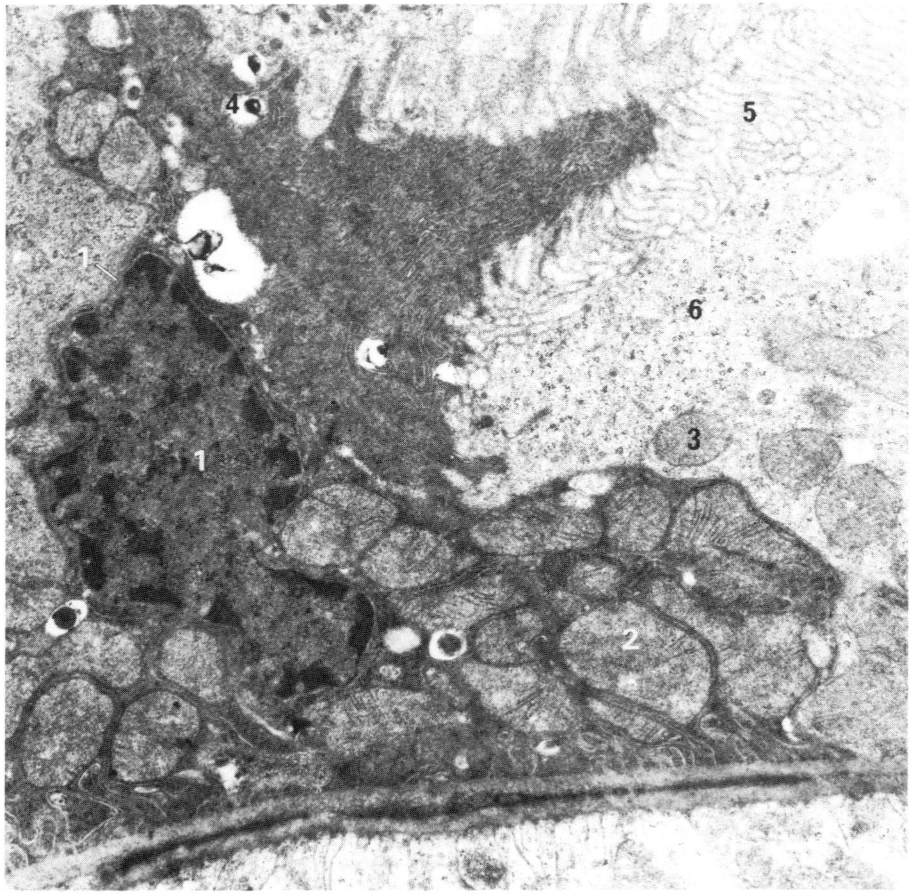

1. Nucleus showing invagination, increased karyoplasmic
density and marginal chromatinization. 2. Swollen and
disrupted mitochondria. 3. Normal mitochondria.
4. Autophagic vesicles. 5. Microvillus brush border.
6. Normal electron density of proximal tubule cell.

Figure 1   Glomerular damage in cadmium contaminated
           kidney from *S. araneus* x12000

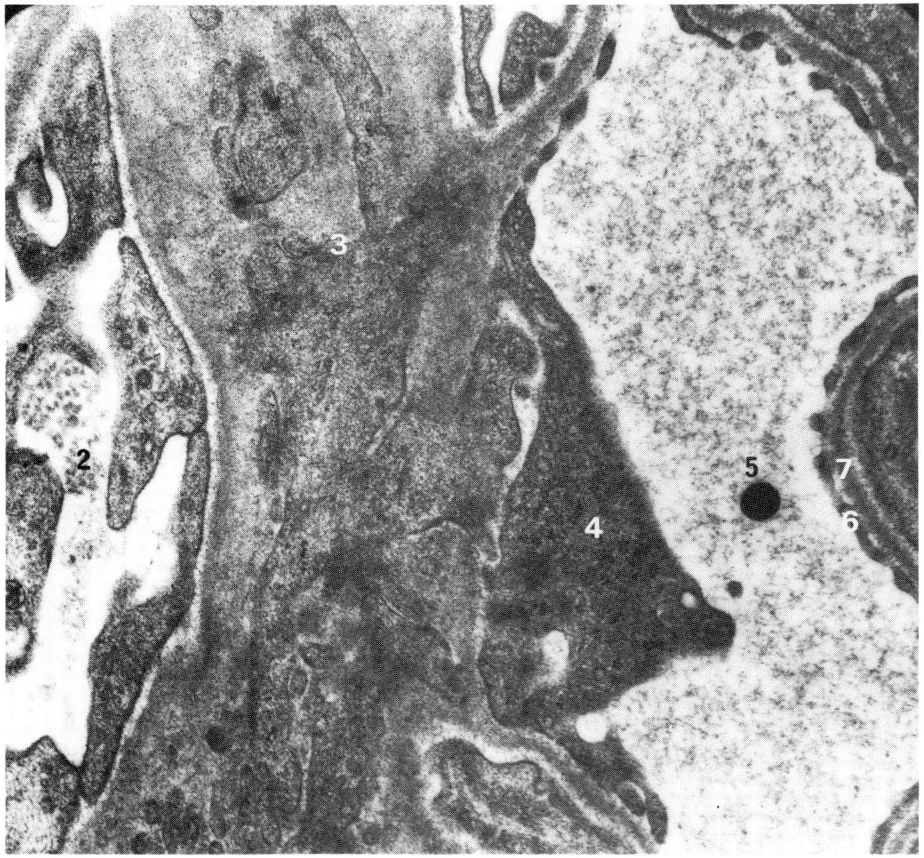

1. Fusion of epithelial podocytic processes.
2. Fibrin in podocytic space. 3. Proliferation
of basement membrane material. 4. Hypertrophy of
endothelial cell. 5. Serum proteins. 6. Normal
endothelial cell thickness. 7. Normal basement
membrane thickness.

Figure 4 Sinusoidal damage and inflammatory cell
invasion in cadmium contaminated *S. araneus*
liver x2000

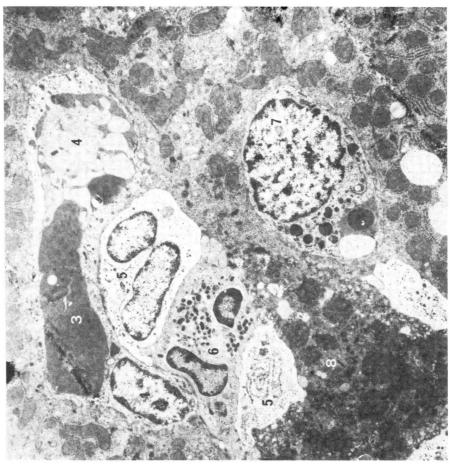

1. Endothelial cell nucleus. 2. Hypertrophy of
endothelium. 3. Red blood cells packing capillary
sinus. 4. Normoblast. 5. Macrophage cells.
6. Neutrophil cell. 7. Kupffer cell. 8. Hepatocytes.

Figure 3 Hepatocyte with cadmium inclusion body
from cadmium contaminated *S. araneus*
x5000

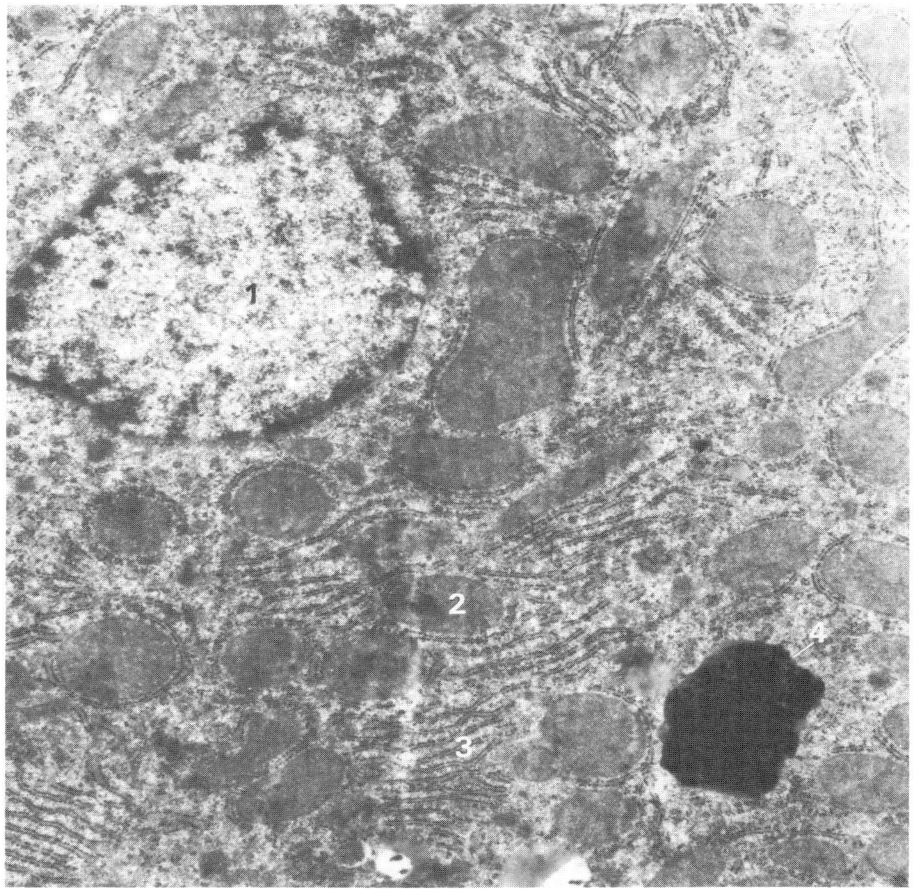

1. Nucleus. 2. Undisrupted mitochondria.
3. Undisrupted rough endoplasmic reticulum.
4. Cadmium inclusion body.

hepatocyte damage was most severe in juvenile *S. araneus*. Whether the greater disruption of liver function in juvenile animals compared to adults represents a form of tolerance to the high cadmium burden in the latter remains unclear.

CONCLUSION

S. araneus from contaminated grasslands shows significant accumulation of copper and cadmium in the liver and kidney target organs. Copper concentrations are approximately double control values, in contrast to cadmium which shows prolific accumulation up to 100 times that of control values, in the adult liver. Electron microscope examination has revealed widespread liver and kidney lesions, the frequency of which increases in line with cadmium concentration. Kidneys showed lesions in both glomerular and proximal tubule cells, but damage was not so severe as that previously described for wild seabirds. Livers showed both hepatocyte and sinusoidal damage. Hepatocytes of adult *S. araneus* contained cadmium inclusion bodies in the cytoplasm. The presence of these hitherto undescribed inclusion bodies was correlated with reduced disruption of RER, SER and mitochondria in hepatocytes. Severe sinusoidal damage was evident with the lumens packed by inflammatory, plasma and Kupffer cells. Such extensive tissue damage may well provide the basis of selection pressures resulting in altered gene pools in contaminated populations. The end point of such continued selection could be the evolution of metal tolerant ecotypes.

REFERENCES

EVANS, G.W., MAJOR, P.F. & CORNATZER, W.E. 1970. Mechanism for cadmium and zinc antagonism of copper metabolism. *Biochem. biophys. Res. Commun.*, 40, 1142-1148.

HUNTER, B.A. & JOHNSON, M.S. 1982. Food chain relationships of copper and cadmium in contaminated grassland ecosystems. *Oikos*, 38, 108-117.

HUNTER, B.A., JOHNSON, M.A., THOMPSON, D.J. & HOLDEN, H. 1981. Age accumulation of copper and cadmium in wild populations of small mammals. In: *Proc. int. Conf. Heavy Metals in the Environment, Amsterdam, 1981*. Edinburgh: CEP Consultants.

NICHOLSON, J.K. & OSBORN, D. 1983. Kidney lesions in pelagic seabirds with high tissue levels of cadmium and mercury. *J. Zool.*, 200, 99-118.

PAGE, A.R. & GOOD, R.A. 1960. Plasma cell hepatitis, with special attention to steroid therapy. *Am. J. Dis. Child.*, 99, 288-314.

UNDERWOOD, E.J. 1977. *Trace elements in human and animal nutrition*. New York: Academic Press.

LOCALISATION OF METALS IN MUSSEL KIDNEY

B J S PIRIE
Institute of Marine Biochemistry, St Fittick's Road, Aberdeen AB1 3RA

ABSTRACT

The mussel kidney is the major storage organ of various metal pollutants. The metals are stored in membrane-bound insoluble granules which occupy 20% of the cell volume. This mechanism immobilizes the metal pollutants and detoxifies them by isolating them from the surrounding cytoplasm. The granules are residual bodies containing inorganic constituents and metals and a pigmented organic component with the properties of lipofuscin.

INTRODUCTION

The common seawater mussel, *Mytilus edulis*, is extensively used as a monitor of metal pollution in estuarine and marine environments, because uptake is dependent on seawater concentration although also affected by metal speciation. Along with other bivalves, *Mytilus* can concentrate the metals in excess of their environmental levels, and high concentrations of cadmium, zinc, lead and iron accumulate in the kidneys. Mussels experimentally exposed to radioactive zinc for 8 days followed by non-radioactive zinc showed continued increase in the radioactivity in the kidney but a decrease in other tissues after the eighth day, indicating transfer to the kidney which acts as a storage site (George & Pirie 1980). Despite a vast amount of concentration data in the literature, little is known of the actual metal metabolism and its relevance to our understanding of pollution research. Therefore, this paper first relates the structure and function of the *Mytilus* excretory system and, second, demonstrates how and where the metals are localised within the kidney.

RESULTS AND DISCUSSION

Structure of the excretory system

The *Mytilus* excretory system comprises a dorsally placed heart consisting of 3 chambers: a ventricle (through which passes the rectum), and paired lateral auricles. Brownish pericardial glands cover the auricles and extend down the afferent oblique veins. These organs are enclosed by a thin-walled pericardium and the base of the pericardial cavity projects ventrally as 2 horns which enclose each of the afferent oblique veins to form the renopericardial canals which enter the kidneys. The dark-brown coloured kidneys lie dorsally along the gill axis. Electron microscopy reveals that the cells of the auricle, pericardial gland and afferent oblique veins (Figure 1) have features characteristic of podocytes and contain numerous dense inclusions which impart the brown pigmentation to the tissue (Pirie & George 1979;

Pirie 1982). The presence of podocytes suggests that these are areas where ultra-filtration can occur. The ultra-filtrate produced in the pericardial cavity is conveyed to the kidneys via the ciliated renopericardial canals.

The kidneys can be regarded as coelomoducts with lobules ending in acini and bathed in the blood of the surrounding sinus. The columnar kidney cells (Figure 2) have an infolded basal membrane and are packed with electron-dense granules occupying at least 20% of the cell volume (Pirie & George 1979; Pirie 1982). Thus, granules are present in various parts of the excretory system but investigation of their chemical composition will reveal differences, particularly in metal content.

Transport of metals

Uptake of metal pollutants in seawater is via the gills, mantle and gut. They are transported from the gills and gut via the haemolymph in granular amoebocytes (Figure 1) or, as in the case of zinc, as a high molecular weight complex in the cell free haemolymph (George & Pirie 1980) to the kidney. Although the kidney contains the highest concentrations of iron, zinc, cadmium and lead, the percentage in the kidney of the body total varies from 70% of the total lead to 40% of the other metals. Therefore, different metabolic processes and storage mechanisms may be involved for each metal.

Localisation of metals

Animals from non-polluted natural environments and animals experimentally exposed to zinc, cadmium and lead in seawater were examined by electron probe X-ray microanalysis. Cryo-sections of pericardial gland and kidney tissue were used as conventional fixation methods have been shown to induce movement and loss of diffusible substances. X-ray microanalysis reveals that the highest concentrations of metals in the cells are localised within the membrane-bound electron-dense granules (Table 1). Iron is found in the granules of both pericardial gland and kidney of all animals. However, zinc, cadmium and lead are only detectable in the kidney granules. The absence of these metals from the pericardial gland may reflect the function of podocytes which would prevent the efflux of ions bound to high molecular weight compounds from the blood, and it is known that zinc exists in such a form in mussel haemolymph (George & Pirie 1980). The kidney granules mostly display an homogeneous matrix, but their composition is variable as indicated by the relatively large standard errors (Table 1). In addition to these metals, they also contain high concentrations of sulphur, chlorine and potassium. Additional perfectly spheroidal granules are occasionally observed in zinc exposed animals and these contain zinc, calcium and phosphorus, suggesting a purely mineral granule derived from a separate biochemical pathway. In lead exposed animals, the lead deposits are scattered throughout the granule matrix and produce a speckled appearance in the electron microscope. It is interesting that 85% of the cadmium is found to be associated with granules by X-ray microanalysis of cryo-sections (Figure 3). However, after cell fractionation which actually ruptures lysosomes, most of the cadmium is found in the cytoplasmic fraction. It is therefore possible that a metallothionein-like protein

TABLE 1   Elemental distribution in cryo-section of lead exposed
           mussel kidney

           Areas ca 30 nm diameter were analysed by electron probe
           X-ray microanalysis. Elemental concentrations are given
           as mass fractions (P-b/W x detector efficiency) ± SEM

Element	Tissue area			
	Granules	Nucleus	Mitochondria	Cytoplasm
P	5.0 ± 2.8	5.8	3.3	13.6 ± 2.7
S	41.6 ± 6.5	14.0	13.8	9.5 ± 2.0
Cl	5.1 ± 1.6	6.2	9.5	9.4 ± 2.3
K	3.9 ± 1.9	6.8	5.4	5.8 ± 1.6
Ca	0.5 ± 0.3	0.2	0.5	0.8 ± 0.5
Fe	9.2 ± 9.9	1.0	1.4	0.1 ± 0.1
Zn	1.9 ± 1.8	0.7	0.7	0.5 ± 0.1
Pb	15.6 ± 6.5	1.8	0.6	1.0 ± 1.3
Areas analysed	30	3	2	5

may be associated with the granules in mussel kidney (George & Pirie
1979). While the granules also contain other inorganics such as
phosphorus, sulphur, chlorine, potassium and calcium, they have, in
addition, a pigmented organic component. Investigations using histo-
chemistry and spectroscopy demonstrate that the organic component of
the granules has the properties of lipofuscin (George *et al.* 1982;
Pirie 1982). Enzymic evidence suggests that the lipofuscin is present
as a result of lysosomal degradation and peroxidation of cellular
membranes. The granules may, therefore, be regarded as tertiary
lysosomes or residual bodies.

    The metal pollutants are thus packaged in a membrane-bound granule
and effectively detoxified. However, examination of the apical surface
of the kidney cells reveals exocytosis mechanisms permitting slow turnover of single granules and portions of cytoplasm, which are observed
in the lumen of the kidney and excretory pore, resulting in the excretion
of a urine containing particulate material (Pirie 1982).

Figure 1   Electron micrograph of the wall of the afferent oblique
           vein showing elongated podocytes lying on a network of
           collagen fibres containing muscle cells. Granular
           amoebocytes are present in the blood sinus x3700

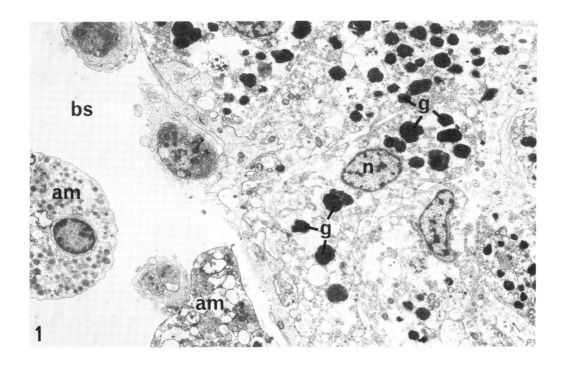

Figure 2   Electron micrograph of transverse section through a kidney
           tubule showing the columnar cells containing many electron-
           dense membrane-limited granules x3000

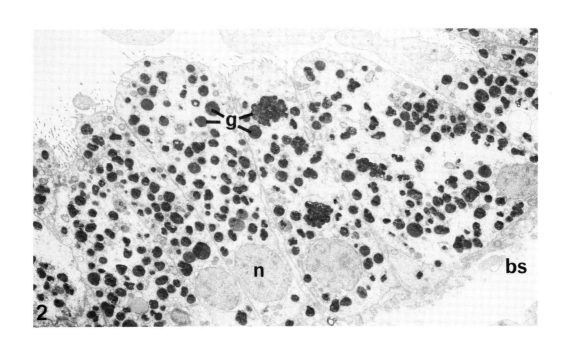

Figure 3   Spectrum of single granule from cryo-section of kidney
of cadmium exposed mussel produced by electron probe
X-ray microanalysis

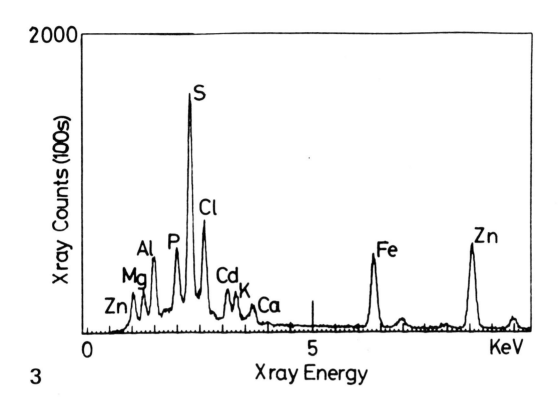

REFERENCES

GEORGE, S.G. & PIRIE, B.J.S. 1979. The occurrence of cadmium in subcellular particles in the kidney of the marine mussel, *Mytilus
edulis*, exposed to cadmium. *Biochim. biophys. Acta*, 580, 234-244.

GEORGE, S.G. & PIRIE, B.J.S. 1980. Metabolism of zinc in the mussel,
Mytilus edulis (L.): a combined ultrastructural and biochemical
study. *J. mar. biol. Ass. UK*, 60, 575-590.

GEORGE, S.G., COOMBS, T.L. & PIRIE, B.J.S. 1982. Characterization of
metal-containing granules from the kidney of the common mussel,
Mytilus edulis. *Biochim. biophys. Acta*, 716, 61-71.

PIRIE, B.J.S. 1982. *An ultrastructural and histochemical study of
the excretory system of the common mussel*, Mytilus edulis (L.).
M.Phil. thesis, Open University.

PIRIE, B.J.S. & GEORGE, S.G. 1979. Ultrastructure of the heart and
excretory system of *Mytilus edulis* (L.). *J. mar. biol. Ass. UK*,
59, 819-829.

THE EFFECTS OF CADMIUM ON THE CONCENTRATIONS OF SOME NATURALLY OCCURRING INTRACELLULAR ELEMENTS IN THE KIDNEY

MARION D KENDALL*, J K NICHOLSON** and ALICE WARLEY***
*Department of Anatomy, St Thomas's Hospital Medical School, London SE1 7EH
**Toxicology Unit, School of Pharmacy, London WC1E 7HX
***Department of Physiology, University of Cambridge, Cambridge

ABSTRACT

Cadmium induced nephrotoxicity is one of the most important serious consequences of exposure to the metal in both man and animals (see Bremner 1978 and Webb 1979 for reviews). Lesions may be seen in the proximal tubules by both light and electron microscopy at quite low cadmium burdens, and in mice the damage is visible before the onset of raised proteinuria (Nicholson 1980). The tubule cells become necrotic and contain swollen mitochondria. The damaged cells are eventually shed to the lumen of the nephron. The distal tubules, in contrast, show few, if any, morphological changes but glomerular damage is seen at high cadmium concentrations.

This paper describes the effects of cadmium administration on the concentrations of some naturally occurring elements in the proximal and distal tubules of mice, as measured by X-ray microanalysis.

MATERIAL AND METHODS

The metal was administered to A_2G Swiss mice as 2 subcutaneous injections of 0.7 µMoles of $CdCl_2$ given 3 days apart. Sampling 3 days later was timed to occur when the kidney cortex [Cd] would be about 100 mg/kg dry wt as found in previous AAS studies (Nicholson 1980). This dosing regime induced metallothionein (D Osborn pers. comm.).

The details of the X-ray microanalysis procedure, preparation of standards and instrument calibration are given in full elsewhere (Warley et al. 1983). Briefly, this study was conducted using a Kevex detector fitted to an EMMA-4 microscope and Link 290 multichannel analyser. An accelerating voltage of 60kV was used with a beam current of 4nA. Tiny pieces of kidney cortex were rapidly frozen in liquid N_2 slush before being cut at -65^0C to -70^0C and collected on to nickel grids. After freeze-drying and carbon coating, the sections were analysed for 100 seconds (live time) with an 0.5 µm diameter probe and a magnification of 10 000. Generally 5 samples each of mitochondria and cytoplasm were analysed from the centre of the cells. Quantification was achieved by reference to standards prepared under identical conditions. In addition, some material was prepared by conventional methods for routine electron microscopy.

RESULTS

Conventional electron microscopy confirmed the presence of nephrotoxic lesions in the proximal tubules. Debris from these cells was
observed further down the nephron in the lumen of the distal tubules
(on the same grid squares). The distal tubules were apparently undamaged.

The morphology of the frozen sections was sufficiently good to
identify mitochondria, apical microvilli, nuclei and the basal infoldings
in cells. At a magnification of 10 000, the probe was sited entirely
within a mitochondrion, or in the adjacent cytoplasm.

A full discussion of the results from normal mice with regard to
variation in elemental concentrations in the same cell type but at
different levels of the nephron is given elsewhere (Kendall *et al.*
1983). There were distinct differences between the elemental compositions of both mitochondria and cytoplasm in both proximal and distal
tubules; each could be said to have a characteristic 'fingerprint' of
elements.

The results of the X-ray microanalysis for cadmium dosed mice
compared with normal mice are shown in Table 1. The low values for
cadmium are a reflection of the methodology, as the Lα line of energy
emission was used in preference to the K lines which have absorption
edge energies outside the maximum efficiency range of the detector.
Also, 100 mg/kg cadmium for whole kidney will be an extremely small
amount to detect by X-ray microanalysis (unless it is concentrated
locally). However, cadmium dosed mice had lower elemental
concentrations, particularly Na and K. In the distal tubules, [S] was
increased (highly significantly in the cytoplasm and mitochondria), [Mg]
was raised only in the mitochondria and [P] was lowered in the cytoplasm.
There was no effect on the cytoplasmic elemental levels in the proximal
tubules.

DISCUSSION

This study indicates that there were highly significant alterations
in the concentrations of some elements in mice dosed with cadmium, although
the cadmium itself was not detectable. Previous studies have all used
much greater dosage rates of cadmium (Kawai *et al.* 1977, mouse renal
cortex; Bell *et al.* 1979, rabbit alveolar macrophages; Chassard-Bouchaud
1981, snail digestive glands). Other work has examined animals known
to have high cadmium contents (Hunter *et al.*). Kawai *et al.* (1977)
approached the problem by the use of autoradiography with $^{109}CdCl_2$ and
X-ray microanalysis. They found that the cytoplasm, basement membranes
and mitochondria of the proximal tubules in particular contained cadmium.
This finding is consistent with the view of Foulkes (1974) that two-thirds
of cadmium entering the kidney does so from the peritubular blood supply,
and only one-third is filtered at the glomerulus. Kawai *et al.* (1977)
also noted, but did not quantify, peak retardations of phosphorus
chlorine, and potassium in proximal tubules; their findings were confirmed
here.

L. I. H. E.
THE MARKLAND LIBRARY
STAND PARK RD, LIVERPOOL. L16 9JD

Other studies have concentrated on the analysis of cadmium in
lysosomes or residual bodies which appear to contain quite high
concentrations. In this study we have sought the effect of the
cadmium earlier on in its accumulation in the cells and after
comparatively small doses.

The most widespread effects of cadmium are seen in the proximal
tubule mitochondria where [Na] and [K] are lowered greatly and [Mg],
[P] and [Cl] are also reduced. This finding is consistent with the morphological
appearance of swollen mitochondria in the damaged proximal tubule cells.
The lower elemental concentrations could be the results of alteration
in mitochondrial volume, but for these to occur there must have been an
effect on the ionic equilibrium between the mitochondria and cytoplasm.
Cadmium *in vitro* is known to inhibit Na/ATPase (and probably Mg ATPase)
(Nechay & Saunders 1977; Rifkin 1965; Wald *et al.* 1974) and alterations
in ion pumping could cause mitochondrial volume changes. If the ion
pumping mechanisms are affected in the cell, and because other authors
have found cadmium in the cytoplasm, it is perhaps surprising that the
cytoplasmic concentrations of other elements remained unaltered in the
proximal tubules. It could be more difficult to detect such changes
in large volumes of cell cytoplasm. Interestingly, there were no
changes in [S], or in [Ca] which is known to be finely controlled
inside cells.

The distal tubules present a different picture after cadmium
dosing. There was no detectable effect on the very mobile metals such
as Na and K, despite the fact that the distal tubules may have a greater
Na/K ATPase activity (Doucet *et al.* 1980). Instead, there were highly
significant rises in [S] in both mitochondria and cytoplasm, a marked
increase in mitochondrial [Mg] and a slight decrease in cytoplasmic
[P]. As metallothionein was induced in these experiments, and
metallothionein has a very high S content, it is tempting to speculate
that this rise in [S] is a reflection of an increased metallothionein
content. Banerjee *et al.* (1982) have shown by immunohistochemical
techniques that in control rats the metallothionein is situated in the
distal and collecting tubules in the kidney. It is not known where
metallothionein is produced when it is induced by a cadmium challenge.

Thus, although at first sight it is not surprising that the
proximal tubules are greatly damaged by cadmium, this study raises many
questions. Does damage in the proximal tubules mean that there is
insufficient metallothionein present to take up the cadmium, or does
damage occur too rapidly for metallothionein to be induced? Similarly,
is the distal tubule not damaged because of its metallothionein content,
or because it can rapidly synthesize metallothionein? These arguments
presuppose that the metallothionein acts to bind the cadmium and render
it non-toxic - is this true? Clearly, the raised [S] in the distal
tubule prompts much more work in this area.

CONCLUSIONS

X-ray microanalysis is a sensitive method for the study of the
effects of metals such as cadmium. This work suggests that, of the
proximal and distal tubules, the principal organelles affected by the

presence of cadmium are the proximal tubule mitochondria. The effect is to lower elemental concentrations of sodium and potassium in particular. An important rise in [S] in the distal tubule cytoplasm and mitochondria was observed and it is postulated that this might be associated with the induction of metallothionein after a cadmium challenge (Bremner 1982).

ACKNOWLEDGEMENTS

We are grateful for financial support from NERC for JKN and from the MRC for AW; to Dr T C Appleton for making the microanalysis facilities available to us; to Mr A Hockaday for technical assistance; and to the St Thomas's Hospital Research (Endowment) Fund for travel and incidental expenses.

REFERENCES

BANERJEE, D., ONOSAKA, S. & CHERIAN, M.G. 1982. Immunohistochemical localisation of metallothionein in the cell nucleus and cytoplasm of rat liver and kidney. *Toxicology*, 24, 95-105.

BELL, S.W., MASTERS, S.K., INGRAM, P., WATERS, M. & SHELBURNE, J.D. 1979. Ultrastructure and X-ray microanalysis of macrophages exposed to cadmium chloride. *Scanning Electron Microsc. (SEM)*, 1979/III, 111-121.

BREMNER, I. 1978. Cadmium toxicity. *Wld Rev. Nutr. Diet*, 32, 165-197.

BREMNER, I. 1982. The nature and function of metallothionein. In: *Trace element metabolism in man and animals*, edited by J.M. Gawthorne, J. McC. Howell & C.L. White, 637-644. Berlin: Springer Verlag.

CHASSARD-BOUCHAUD, C. 1981. Rôle des lysosomes dans le phénomène de concentration du cadmium. Microanalyse par spectrographie des rayons X. *C. r. Acad. Sci. Paris*, 293, Sér. III, 261-265.

DOUCET, A., MOREL, F. & KATZ, A.I. 1980. Microdetermination of Na-K-ATPase in single tubules: its application for the localisation of physiological processes in the nephron. *Int. J. Biochem.*, 12, 17-52.

FOULKES, E.C. 1974. Excretion and retention of cadmium, zinc and mercury by rabbit kidney. *Am. J. Physiol.*, 227, 1356-1360.

KAWAI, K., KYONO, H., SAKAI, T. & MURAKAMI, M. 1977. Localisation of cadmium in the renal cortex. In: *Clinical chemistry and chemical toxicology of metals*, edited by S.S. Brown, 113-118. Amsterdam: Elsevier/North Holland Biomedical Press.

KENDALL, M.D., WARLEY, A., NICHOLSON, J.K. & APPLETON, T.C. 1983. X-ray microanalysis of proximal and distal tubule cells in the mouse kidney, and the influence of cadmium on the concentration of natural intracellular elements. *J. Cell Sci.* (in press).

NECHAY, B.R. & SAUNDERS, J.P. 1977. Inhibition of renal adenosine triphosphatase by cadmium. *J. Pharmac. exp. Ther.*, 200, 623-629.

NICHOLSON, J.K. 1980. *Studies on the kidney in relation to the pharmacodynamics of zinc, cadmium and mercury*. PhD thesis, University of London.

RIFKIN, R.J. 1965. *In vitro* inhibition of $Na^+ - K^+$ and Mg^{2+} ATPase by mono, di and trivalent cations. *Proc. Soc. exp. Biol. Med.*, 120, 802-804.

WALD, H., GUTMAN, Y. & CZACKES, W. 1974. Comparison of microthermal ATPase in the cortex, medulla and papille of the rat kidney. *Pflügers Arch. ges. Physiol.*, 352, 47-59.

WARLEY, A., STEPHEN, J., HOCKADAY, A. & APPLETON, T.C. 1983. X-ray microanalysis of HeLa S3. I: Instrumental calibration and analysis of randomly growing cultures. *J. Cell Sci.*, 60, 217-229.

WEBB, M. 1979. *The chemistry, biochemistry and biology of cadmium.* (Topics in Environmental Health 2.) Amsterdam: Elsevier/North Holland Biomedical Press.

TABLE 1 The mean concentration (mmol/kg freeze dried mass) of elements in the cytoplasm and mitochondria of proximal and distal tubules

	Na	Mg	P	S	Cl	K	Ca	Cd	Nos
PROXIMAL TUBULES									
Mitochondria									
Normal	221(±23)	58(±7)	516(±40)	251(±18)	243(±27)	311(±26)	10(±9)	-5(±1)	24
Cd dosed	121(±9)	34(±6)	410(±28)	242(±20)	173(±10)	217(±18)	21(±7)	-1(±1)	29
t values	4.31	2.67	2.23	0.31	2.64	3.07	-0.91	-2.20	df = 51
Effect	***↓	*↓	*↓	—	*↓	**↓	—	—	
Cytoplasm									
Normal	109(±14)	26(±8)	166(±16)	95(±10)	74(±11)	124(±13)	-4(±13)	-1(±2)	21
Cd dosed	82(±14)	14(±6)	170(±13)	93(±7)	56(±5)	125(±9)	12(±6)	<1(±1)	30
t values	1.36	1.18	-0.19	0.13	1.75	-0.04	-1.25	-0.53	df = 49
Effect	—	—	—	—	—	—	—	—	
DISTAL TUBULES									
Mitochondria									
Normal	89(±11)	25(±6)	509(±24)	156(±12)	252(±18)	392(±17)	18(±10)	<-1(±2)	24
Cd dosed	109(±12)	55(±10)	526(±33)	334(±25)	265(±32)	368(±27)	39(±11)	<1(±3)	18
t values	-1.24	-2.59	-0.42	-6.97	-0.40	0.79	-1.39	-0.16	df = 48
Effect	—	**↑	—	***↑	—	—	—	—	
Cytoplasm									
Normal	73(±13)	12(±6)	256(±23)	58(±4)	77(±6)	222(±20)	6(±10)	-2(±2)	21
Cd dosed	84(±12)	22(±9)	188(±20)	126(±12)	84(±15)	181(±15)	-6(±8)	-2(±2)	23
t values	-0.60	-0.97	2.20	-5.54	-0.47	1.62	0.89	-0.21	df = 45
Effect	—	—	*↓	***↑	—	—	—	—	

$* = p < 0.1$; $** = p < 0.01$; $*** = p < 0.001$

III. METALLOTHIONEIN

FUNCTIONS OF METALLOTHIONEIN

K CAIN
Toxicology Unit, Medical Research Council Laboratories, Carshalton, Surrey SM5 4EF

ABSTRACT

Metallothionein is a unique, inducible metalloprotein with a very high affinity for cadmium, mercury, zinc, copper and other metals. Thionein synthesis in response to an inducing metal can provide a means of heavy metal detoxification or, alternatively, a homeostatic mechanism for controlling essential metal metabolism. This paper reviews some of the evidence for these proposed functions.

REVIEW

The subject of this paper covers a very large topic and it is obviously impossible to cover every aspect in sufficient detail. Thus, of necessity, this is really an overview, rather than a comprehensive review; hopefully, the bibliography will enable more interested readers to expand their knowledge. The discovery of metallothionein is a direct result of studies on cadmium which is a natural component of the environment. Cadmium ion is toxic and the metal once absorbed is retained with very little excretion; as a result, even at low-level exposures, long-lived animal species accumulate significant body burdens in their life-times. The main target organs are the liver and the kidneys, in which it has been shown by subcellular distribution studies (eg Sabbioni *et al.* 1978) that over 80% of the tissue cadmium is located in the cell cytosol. The explanation for this subcellular distribution was provided earlier by Margoshes and Vallee (1957), who showed that the soluble cadmium was bound to a metalloprotein now known as a metallothionein. Since this discovery, a great deal of work has been carried out on the properties and function of metallothionein. To summarise, metallothionein is a small (6000 MW) protein which can be easily isolated from the cytosol by gel filtration. This crude metallothionein preparation can be further purified by anion-exchange chromatography into at least 2 iso-metallothioneins (Mt1 and Mt2). All metallothioneins are deficient in aromatic amino acids and contain appreciable amounts of lysine and serine, in addition to the 20 cysteinyl residues per molecule. The primary structure is highly conserved, and thus the location of the cysteines (all of which are involved in cation binding), for example in rat liver thionein is similar to the mouse, rabbit and human forms. All of the cysteinyl SH groups are involved in metal binding and the protein has a high affinity for Cd^{2+}, Zn^{2+}, Cu^{2+} and Hg^{2+}. Each metal centre (in the case of Cd^{2+}, this is 7 per protein molecule) has a negative charge. However, the total charge in the protein is also influenced by other charged amino acids, the composition of which (and, as a result, the total charge of the protein) varies in different species.

Cd-metallothionein is synthesized in the liver, kidney and intestine, and in the case of the liver is located in the parenchymal rather than the non-parenchymal (Kupffer) cells (Cain & Skilleter 1981). This is due to the greater uptake of the metal into the former cells and not to an increased ability to synthesize metallothionein (Cain & Skilleter 1983). It is therefore likely that the preferential synthesis of metallothionein in the liver, kidney and intestine is a reflection of the rate of metal uptake into these tissues, rather than a superior ability to synthesize the metalloprotein. In this context, it should be stressed that the long biological half-life of cadmium in these organs cannot be explained by the high affinity of the metal for the protein, as there is a continual turnover of the protein moiety. Thus, the metallothionein is continually degraded and the released Cd^{2+} rebound by the newly synthesized thionein (Cain & Holt 1979). As illustrated in Table 1, the protein turnover times ($T\frac{1}{2}$) of the isometallothioneins are different and vary according to the metal composition.

TABLE 1   Turnover times ($T\frac{1}{2}$) and metal ratios of hepatic cadmium-thionein

Data taken from Cain and Holt (1979)

	Low Cd^{2+} dose			High Cd^{2+} dose		
	$T\frac{1}{2}$ (days)	Cd/Zn (g atom/ g atom)	Cd/Cu (g atom/ g atom)	$T\frac{1}{2}$ (days)	Cd/Zn (g atom/ g atom)	Cd/Cu (g atom/ g atom)
Mt1	2.2	0.63	4.6	3.1	1.5	9.42
Mt2	3.5	0.57	9.42	5.1	1.31	56.9

The normal, unexposed animal has little or no metallothionein in the liver, kidneys and intestine. However, as shown by Piscator (1964), chronic exposure to cadmium leads to the progressive accumulation in the liver and kidneys of not only the metal but also the metallothionein. These findings led Piscator to propose that the protein was synthesized in response to the cadmium as a 'protective detoxifying mechanism'. Since this original discovery, a considerable amount of research has been carried out (see Webb 1979 for review) and the evidence demonstrates that the theory works to a certain extent. However, it is apparent that the protective effects of thionein synthesis against cadmium toxicity have some limitations, as, at chronic exposure to cadmium, kidney failure is the prime symptom and occurs at a concentration of about 200 mg Cd^{2+}/kg kidney. At this level of exposure, a considerable percentage of the tissue cadmium is bound to thionein and it is likely that it is the non-thionein bound cadmium which is the toxic species. In this respect, Nomiyama and Nomiyama (1981) have shown that this is not a fixed percentage of the total tissue cadmium and can vary with species and dosing. An alternative explanation is that, at high levels of exposure, cadmium uptake and the amount of the metal released by thionein degradation exceed the rate of thionein synthesis and, as a result, the percentage of non-thionein bound cadmium increases. This suggestion is borne out by the feeding experiment shown in Figure 1, in which the percentage of thionein bound cadmium decreases with respect to the time spent on the diet.

Figure 1   Total (●) and thionein (○) cadmium in the kidneys of rats
          fed on a 100 mg/kg cadmium diet

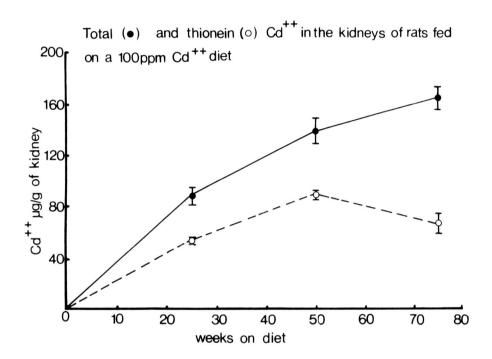

In any event, it is apparent that the protective role of thionein is only effective at low levels of chronic exposure. This is illustrated in the experimental animal by the simultaneous injections of $CdCl_2$ and excess cysteine (Murakami & Webb 1981) which produce a renal concentration of 20 µg Cd^{2+}/g within one hour of dosing. Although synthesis of thionein begins within 2-7 hours of injection, the animals die from renal failure. Nephrotoxicity, therefore, can result from a small renal burden of cadmium, providing its accumulation is rapid and maximal before thionein synthesis is initiated.

The protective function of thionein against cadmium toxicity is well established and generally accepted. However, there is now an increasing amount of evidence to suggest that this detoxification function may be nothing more than the fortuitous interaction of cadmium and other heavy metals with the normal homeostatic mechanisms for zinc and copper. Important developments in this concept were provided by the work of Bremner who showed that there was a critical level (30 ppm) for zinc in the liver (Bremner *et al*. 1973) above which thionein synthesis was stimulated, and Cousins who demonstrated that the intestinal mucosal cells also synthesized thionein in response to elevated zinc levels (Richards & Cousins 1977). From these and subsequent studies, Cousins has accumulated evidence that the zinc status of the adult animal is controlled by metallothionein synthesis in the liver and intestine. In the proposed scheme (Cousins 1979), the intracellular zinc content of the mucosal cell consists of a mobile pool in equilibrium with zinc-metalloprotein and other non-exchangeable zinc proteins. Elevated zinc levels in the diet lead to an increased absorption across the mucosal cell membrane and an increased metallothionein synthesis which serves to limit the transfer of zinc to the plasma. The model is reversible and increased plasma levels of the metal lead to an increased metallothionein concentration in the mucosal cell, which may be excreted by desquamation. Similar models have been advanced to explain the regulation of copper and cadmium absorption (see Webb & Cain 1982 for a fuller discussion). In the case of the liver, the intracellular zinc pool of the hepatocyte is regulated by the plasma zinc concentration. A rise in the latter leads to an elevated zinc level in the cytosol of the liver and a concomitant increase in thionein synthesis. The model also envisages an additional control point, in that thionein synthesis may be induced by glucocorticoid hormones which have been shown to stimulate zinc uptake and zinc-thionein synthesis in isolated hepatocytes (Karin *et al*. 1980) and HeLa cells (Karin & Herschman 1981). This may explain the increases in thionein levels seen after various physiological stresses (Webb & Cain 1982), as these invariably involve changes in the levels of the adrenal and pituitary hormones. Clearly, zinc-thionein synthesis responds rapidly to variations in zinc levels and this satisfies an important requirement for a homeostatic mechanism. The metalloprotein is a convenient form of storing zinc, but it should also be capable of supplying the cation to appropriate metabolic processes. In support of this, *in vitro* experiments have shown that the metallothionein can be used to reactivate zinc-depleted apo-enzymes by cation donation (Udom & Brady 1980). As yet there is no evidence that this occurs *in vivo* and it is possible that the metal is made available by protein degradation. Unlike Cd-thionein, the zinc protein has a $T\frac{1}{2}$ of only 18 hours and its degradation is accompanied by the release of the cation which is not rebound by newly synthesized thionein (Feldman & Cousins 1978).

If metal storage and donation are the primary functions of metallothionein, it should be possible to correlate thionein synthesis with metabolic demand. Such a situation is seen in the regenerating liver which after partial hepatectomy has an absolute requirement for zinc. The metal is presumably incorporated into newly synthesized zinc-dependent enzymes like DNA-polymerase, the concentrations of which are known to increase rapidly during regeneration. Virtually all of the accumulated zinc is bound initially to thionein, the synthesis of which precedes the regenerative process (Ohtake & Koga 1979). Interestingly, although both isometallothioneins are synthesized, there is a much greater accumulation of Mt2 (Cain & Griffiths unpublished results) which may reflect a specific role for Mt2 in regeneration. Alternatively, cellular protein degradation is markedly reduced in the regenerating liver, and the apparently preferential synthesis of Mt2 may be nothing more than a fortuitous consequence of these metabolic changes.

Metallothionein may also be involved in cell growth in the foetal and newborn animal. Unlike the adult, thionein occurs in high concentrations in the liver of many foetal and newborn animals. In the rat, for example, hepatic zinc and thionein levels rise rapidly before birth and reach a maximum at 2 days post-partum (Mason *et al.* 1981a). Thereafter, as the liver grows, the total zinc concentration steadily decreases to the adult level of 30 mg/kg at 26 days of age, at which time the concentration of metallothionein zinc is only 0.5 mg/kg. However, the total content of thionein bound zinc remains relatively constant until the 16th day, after which it decreases. It has been proposed that the thionein bound zinc is a 'reserve' for the developing liver. The situation in the liver is also paralleled in the intestine which appears to have its own regulatory mechanism for copper absorption. Thus, at birth, the copper concentration in the rat intestine is 17 mg/kg and this increases to approximately 140 mg/kg (Mason *et al.* 1981b) 2 days post-partum as at least 60% of the ingested copper is retained. Most of the copper is found in the cytosol of the intestinal mucosa as a heterogeneous polydisperse metal complex, one component of which may be metallothionein (Mason *et al.* 1981b). The concentration of the complex remains relatively constant until the 13-15th day after birth when there is a sudden loss of copper, and at 21 days the concentration and distribution are similar to that of the adult. This loss of copper is coincident with the maturation of the immature intestine. Thus, in the newborn rat, there exist appropriate mechanisms to control copper and zinc levels. As discussed previously (Webb & Cain 1982; Cain & Webb 1983), these mechanisms may differ from species to species. The newborn Syrian hamster, for example, does not have an intestinal copper complex, and zinc and copper homeostasis are apparently regulated solely by the levels of hepatic thionein which binds both metals (Bakka & Webb 1981). The intestinal complex and hepatic thionein can also provide the newborn animal with a potential reservoir of immediately available binding sites which can sequester toxic cations like Cd^{2+} and Hg^{2+}, and render them harmless. For example, in the suckling rat, Cd^{2+} absorption is greater than in the adult, but the majority of the metal is retained in the intestine (Kostial *et al.* 1979, 1980), presumably bound to the copper complex (Sasser & Jarboe 1977). Similar experiments have been carried out with mercury (Webb & Holt 1982) which show that retention of the metal in the intestine is greater in the 4 day old suckling rat than in the weanling (21 day old) animal. In this respect, species differences in the gastro-intestinal absorption of cadmium may be related to the maturation of the intestine and the coincident loss of the copper complex (Sasser & Jarboe 1980). In conclusion, the exact function of metallothionein in the foetal and newborn animal is not understood, but obviously it is of some importance. Its importance is perhaps best illustrated by studies in the rat which show that cadmium treatment of the pregnant mother blocks placental transport of zinc, and hence the increase in hepatic zinc-thionein (Bakka *et al.* 1981). At birth the placental block is removed and the liver rapidly accumulates zinc and hepatic zinc-thionein.

REFERENCES

BAKKA, A. & WEBB, M. 1981. Metabolism of zinc and copper in the neonate: changes in the concentrations and contents of thionein-bound Zn and Cu with age in the livers of the newborn of various mammalian species. *Biochem. Pharmac.*, 30, 721-755.

BAKKA, A., SAMARAWICKRAMA, G.P. & WEBB, M. 1981. Metabolism of zinc and copper in the Neonhet; effect of cadmium administration during late gestation in the rat on zinc and copper metabolism of the new born. *Chemico-biol. Interactions*, 34, 161-171.

BREMNER, I., DAVIES, N.T. & MILLS, C.F. 1973. The effect of zinc deficiency and food restriction on hepatic zinc proteins in the rat. *Trans. Biochem. Soc.*, 1, 982-985.

CAIN, K. & HOLT, D.E. 1979. Metallothionein degradation: metal composition as a controlling factor. *Chemico-biol. Interactions*, 28, 91-106.

CAIN, K. & SKILLETER, D.N. 1981. Selective uptake of cadmium by the parenchymal cells of the liver. *Biochem. J.*, 188, 285-288.

CAIN, K. & SKILLETER, D.N. 1983. Comparison of Cd-metallothionein synthesis in parenchymal and non-parenchymal rat liver cells. *Biochem. J.*, 210, 769-773.

CAIN, K. & WEBB, M. 1983. Metallothionein and its relationship to the toxicity of cadmium and other metals in the young. In: *Proc. int. Health Evaluation of Heavy Metals in Infant Formula and Junior Food, Berlin (West), 1981*, edited by E.H.F. Schmidt & A.G. Hilderbrandt, 105-111. Berlin: Springer-Verlag.

COUSINS, R.J. 1979. Regulation of zinc metabolism in liver and intestine. *Nutr. Rev.*, 37, 97-103.

FELDMAN, S.L. & COUSINS, R.J. 1978. Degradation of hepatic zinc-thionein after parenteral zinc administration. *Biochem. J.*, 160, 583-588.

KARIN, M., HERSCHMAN, H.R. & WERNSTEIN, D. 1980. Primary induction of metallothionein by dexamethasone in cultured rat hepatocytes. *Biochem. biophys. Res. Commun.*, 92, 1052-1059.

KARIN, M. & HERSCHMAN, H.R. 1981. Induction of metallothionein in HeLa cells by dexamethasone and zinc. *Eur. J. Biochem.*, 113, 267-272.

KOSTIAL, K., KELLO, D., BLANUSA, M., MALJKOVIC, T. & RABAR, I. 1979. Influence of some factors on cadmium pharmacokinetics and toxicity. *Environ. Health Perspect.*, 28, 89-95.

KOSTIAL, K., RABAR, I., BLANUSA, M. & CIGANOVIC, M. 1980. Influence of trace elements on cadmium and mercury absorption in sucklings. *Bull. environ. Contam. Toxicol.*, 25, 436-440.

MARGOSHES, M. & VALLEE, B.L. 1957. A cadmium protein from equine kidney cortex. *J. Am. Chem. Soc.*, 79, 4813-4814.

MASON, R., BAKKA, A., SAMARAWICKRAMA, G.P. & WEBB, M. 1981a. Metabolism of zinc and copper in the neonate: accumulation and function of (Zn, Cu)-metallothionein in the liver of newborn rat. *Br. J. Nutr.*, 45, 375-389.

MASON, R., BRADY, F.O. & WEBB, M. 1981b. Metabolism of zinc and copper in the neonate: accumulation of copper in the gastro-intestinal tract of the newborn rat. *Br. J. Nutr.*, 45, 391-399.

MURAKAMI, M. & WEBB, M. 1981. A morphological and biochemical study of the effects of L-cysteine on the renal uptake and nephrotoxicity of cadmium. *Br. J. exp. Path.*, 62, 115-130.

NOMIYAMA, K. & NOMIYAMA, H. 1981. Abstract, US-Japan workshop on metallothionein, Cincinnati 23-27 March (1981).

OHTAKE, H. & KOGA, M. 1979. Purification and characterisation of zinc-binding proteins from the liver of the partially hepatectomized rat. *Biochem. J.*, 183, 683-690.

PISCATOR, M. 1964. On cadmium in normal human kidney, together with a report on the isolation of metallothionein from cadmium exposed rabbits. *Nord. hyg. Tidskr*, 45, 76-82.

RICHARDS, M.P. & COUSINS, R.J. 1977. Isolation of an intestinal metallothionein induced by parenteral zinc. *Biochem. biophys. Res. Commun.*, 75, 286-293.

SABBIONI, E., MARAFANTE, E., PIETRA, R., AMARTINI, L. & UBERTALLI, L. 1978. Long term low level exposure experiments by nuclear and radiochemical techniques. In: *Proc. Cadmium Symposium, Jena, August 1977*, 111-116.

SASSER, L.B. & JARBOE, G.E. 1977. Intestinal absorption and retention in neonatal rat. *Toxicol. appl. Pharmacol.*, 41, 423-431.

SASSER, L.B. & JARBOE, G.E. 1980. Intestinal absorption and retention of cadmium in neonatal pigs compared to rats and guinea pigs. *J. Nutr.*, 110, 1641-1647.

UDOM, A. & BRADY, F.O. 1980. Reactivation *in vitro* of zinc-requiring apo-enzymes by rat liver zinc-thionein. *Biochem. J.*, 187, 329-335.

WEBB, M. 1979. *The chemistry, biochemistry and biology of cadmium*. Amsterdam: Elsevier/North Holland.

WEBB, M. & CAIN, K. 1982. Functions of metallothionein. *Biochem. Pharmac.*, 31, 137-142.

WEBB, M. & HOLT, B. 1982. Endogenous metal binding proteins in relation to the differences in absorption and distribution of mercury in new born and adult rats. *Arch. Toxicol.*, 49, 237-245.

THE EFFECT OF MERCAPTOETHANOL ON THE CHROMATOGRAPHIC PROPERTIES OF CRAB (*CANCER PAGURUS*) METALLOTHIONEIN

J OVERNELL

NERC Institute of Marine Biochemistry, St Fittick's Road, Aberdeen AB1 3RA

ABSTRACT

The chromatographic properties of crab metallothionein appear particularly sensitive to oxidation, and this oxidation does not take place when mercaptoethanol is present in the buffers.

INTRODUCTION

The edible crab (*Cancer pagurus*) is able to accumulate cadmium in the hepatopancreas to high levels when growing in apparently unpolluted environments (see Overnell & Trewhella 1979). This cadmium has been found to be bound to the low molecular weight cadmium binding protein metallothionein which is characterized by a high (30%) cysteine content (see Kägi & Nordberg 1979).

Purification of vertebrate metallothionein is generally straightforward: after homogenization and centrifugation (100 000 x g 1 hr), the supernatant is chromatographed on Sephadex G75. The main cadmium containing peak which has an apparent molecular weight of 10 000 is rechromatographed on DEAE cellulose and finally purified by further gel exclusion chromatography.

Crab metallothionein is more difficult to purify. Attempts to use the above procedure failed at the DEAE chromatography step - the cadmium failed to bind. This failure was probably due to proteolysis as a heat treatment of the cytosol (60^0C for 10 min) enabled the material to bind. However, subsequent elution gave a poorly resolved chromatogram (Overnell & Trewhella 1979). In an attempt to improve this isolation, Overnell (1982a) developed an isolation procedure in which the metallothionein was absorbed on to DEAE cellulose from the diluted cytosol and then eluted in a single step. This process achieved a useful preconcentration, but, more importantly, separated metallothionein from all measurable protease activity. Subsequent gel exclusion and ion exchange chromatography were carried out under reduced oxygen tension. (Solutions were gassed with nitrogen or argon and eluted fractions were collected under argon.) Metallothioneins were prepared and characterized, but the resolution of the DEAE chromatograms was still somewhat poor.

Although Porter used mercaptoethanol in 1968 to solubilize a copper protein from neonatal liver, which was subsequently shown to be copper metallothionein, mercaptoethanol has only recently become widely used in the isolation of metallothionein. This renaissance was brought about by the report of Minkel *et al.* (1980) who found that yields of monomeric metallothionein were enhanced when mercaptoethanol was present. Thus, yields of metals in the metallothionein peak in the G75 chromatograms of rat liver cytosol were increased when mercaptoethanol was present in the elution buffer, compared with use of buffers with no added mercaptoethanol, to the following extent: cadmium 86-94%, zinc 25-42%, copper 36-35%.

This paper reports the differences between the chromatograms of crab metallothionein(s) when chromatographed under reducing and non-reducing conditions.

MATERIALS AND METHODS

Crabs were given 2 injections one week apart of 5 mg cadmium per kg live weight and left for 5 months. Two animals were killed and the hepatopancreases were removed and homogenized at $4^{0}C$ with twice their weight of 0.005M mercaptoethanol. The homogenates were centrifuged at 100 000 x g for 1 hour and the filtered cytosols gassed with nitrogen and stored at $-80^{0}C$ until required.

For chromatography under reducing conditions, the cytosols (2 ml) were desalted on a Sephadex G25 column, 5.5 x 1.5 cm (Pharmacia PD10), equilibrated with 0.015M Tris HCl, 0.005M mercaptoethanol, pH 8.6. The product was applied to a column of DEAE cellulose 0.9 x 8.5 cm. This column was prepared from the fines floated off with water from 10 g of Whatman DE52. The column was eluted with 100 ml of buffer gradient from 0.015M Tris HCl, 15 mg/l KCl, 0.005M mercaptoethanol, pH 8.6 to 0.3M Tris HCl, 300 mg/l KCl, 0.005M mercaptoethanol, pH 8.6. The buffer in the mixing chamber was gassed with argon. The Tris concentration in the eluates, which is plotted in the Figures, was determined from the potassium concentrations. It was assumed that there is no interaction of potassium with the ion exchanger.

For chromatography under aerobic conditions, the cytosols (2 ml) were desalted as before, except that mercaptoethanol was not present in the buffer. The product was gently aerated at room temperature for 2 hours and applied to the DEAE cellulose column. This was eluted with the same buffer system as before modified only by the omission of mercaptoethanol. Inert gas was not used.

Metals were determined by direct aspiration into a flame atomic absorption spectrophotometer (Techtron AA5), using standards in hydrochloric acid.

RESULTS AND DISCUSSION

Figures 1A and 1B show the DEAE cellulose chromatograms, obtained under reducing conditions, of hepatopancreas cytosols from 2 individuals (A and B) treated with cadmium.

In separate experiments (Overnell in preparation), it was shown that these metal peaks all elute from Sephadex G75 at a position expected for

Figures 1a, 1b, DEAE cellulose chromatograms of hepatopancreas cytosols
2a, 2b from 2 individual crabs, A and B, injected with cadmium.
Figures 1a and b were obtained under reducing conditions
and Figures 2a and b were obtained under non-reducing
conditions after gentle aeration. Abcissa values are
fraction numbers. Metal concentrations are given in mg/1
(equivalent to µg/ml and ppm)

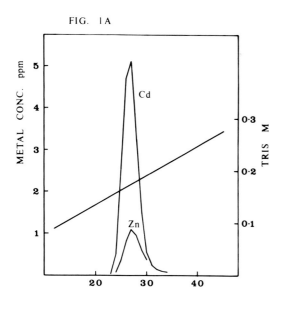

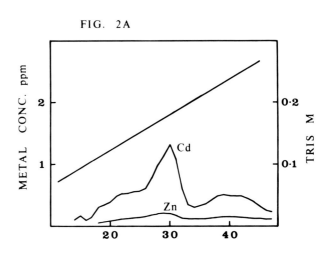

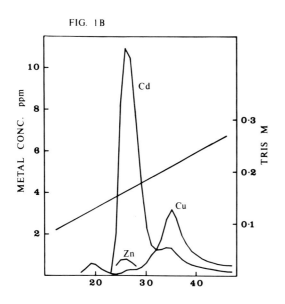

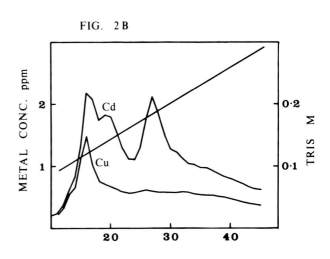

metallothionein, ie with a molecular weight of ca 10 000. Previous work (Overnell 1982b) on the purification of metal binding components from crabs has shown that these metals are bound to metallothionein. In Figure 1A, the cadmium is bound in a single peak which coincides with a peak of zinc. Practically no copper was eluted and its concentration in the cytosol was very low. The cytosolic metal concentrations were copper 4.9, cadmium 28 and zinc 13 mg/l. In Figure 1B, the cadmium is present mainly in a peak at the same position as that of Figure 1A, but the copper is present in 2 separate and different peaks. The cytosolic metal concentrations were copper 44, cadmium 90 and zinc 13 mg/l. The different metal compositions of the different charge forms in Figure 1B will be the subject of a separate communication.

Figures 2A and 2B are the chromatograms of the same 2 samples as above, obtained under aerobic conditions. In Figure 2A, the presence of multiple charge forms of the cadmium is indicated. In Figure 2B, the cadmium is again distributed between multiple charge forms, but the highest concentration of copper now corresponds to one of these. It is suggested here that oxidation brings about the change from Figure 1 to Figure 2 rather than reduction bringing about the change from 2 to 1.

Isolation of vertebrate metallothioneins including fish metallothionein can be carried out under normal aerobic conditions as long as the copper content is not high. Thus, Overnell and Coombs (1979) were able to isolate a cadmium metallothionein from cadmium injected plaice without the need for anaerobic or reducing conditions. Although the metallothionein from crab A is similar to that of the plaice in that it contains mostly cadmium and no copper, yet it is very susceptible to oxidation. The chromatographic pattern illustrated in Figure 2A is similar to that obtained earlier by Overnell (1982a) (as Figure 3A) in the course of purification of crab cadmium metallothionein. In Figure 2A, the main peak appears at a similar although slightly higher Tris concentration to the single peak in Figure 1A and so may well represent the same species. Figure 2B is strikingly similar to that obtained earlier by Overnell (1982b) (as Figure 4) in the course of purification of a crab cadmium, copper metallothionein.

CONCLUSION

The similarity of the aerobic chromatograms presented here and those published earlier suggests that collecting fractions under argon and using argon gassed buffers were not adequate to protect crab metallothionein from changes brought about by oxygen. The net electrical charge on the crab metallothionein molecule appears to be very sensitive to the presence of oxygen in the absence of mercaptoethanol. It thus appears that mercapto-ethanol or other thiol should be present in all buffers used to isolate crab metallothionein.

REFERENCES

KÄGI, J.H.R. & NORDBERG, M. 1979. Metallothionein and other low molecular weight metal-binding proteins. *Experientia Supplementum*, 34. Basel: Birkhäuser Verlag.

MINKEL, D.T., POULSEN, K., WIELGUS, S., SHAW, C.F. & PETERING, D.H. 1980. On the sensitivity of metallothioneins to oxidation during isolation. *Biochem. J.*, 191, 475-485.

OVERNELL, J. 1982a. A method for the isolation of metallothionein from the hepatopancreas of the crab *Cancer pagurus* that minimizes the effect of the tissue proteases. *Comp. Biochem. Physiol.*, 73B, 547-553.

OVERNELL, J. 1982b. Copper metabolism in crabs and metallothionein: *In vivo* effects of copper[II] on soluble hepatopancreas metal binding components in the crab *Cancer pagurus* containing varying amounts of cadmium. *Comp. Biochem. Physiol.*, 73B, 555-564.

OVERNELL, J. & COOMBS, T.L. 1979. Purification and properties of plaice metallothionein, a cadmium binding protein from the liver of the plaice (*Pleuronectes platessa*). *Biochem. J.*, 183, 277-283.

OVERNELL, J. & TREWHELLA, E. 1979. Evidence for the natural occurrence of (cadmium, copper)-metallothionein in the crab *Cancer pagurus*. *Comp. Biochem. Physiol.*, 64C, 69-76.

PORTER, H. 1968. Neonatal hepatic mitochondrocuprein. III. Solubilization of the copper and protein from mitochondria of newborn liver by reduction with mercaptoethanol. *Biochim. biophys. Acta*, 154, 236-238.

THE ISOLATION OF COPPER-METALLOTHIONEIN FROM THE PARTICULATE FRACTIONS OF PIG LIVER

R K MEHRA and I BREMNER
Rowett Research Institute, Bucksburn, Aberdeen AB2 9SB

INTRODUCTION

The incidence of liver dysfunction in copper loaded animals cannot be related to their liver copper content and is probably influenced by factors such as the intracellular distribution of the copper and its molecular association. Histological and electron microscopical examination of copper loaded livers has frequently shown that the metal accumulates in electron-dense granules probably derived from lysosomes (see Bremner 1979a). Subcellular fractionation has confirmed that the copper occurs mainly in particulate fractions, although it often sediments in the nuclear rather than in the lysosomal fractions. Although the binding of cytosolic copper to metallothionein has attracted considerable attention (Bremner 1979b), little is known of how copper is bound in the particulate fractions of the liver. Porter (1971) suggested that the insoluble copper-protein, mitochondrocuprein, in the liver of newborn animals, was a polymerised form of copper-metallothionein. This suggestion was substantiated by Riordan and Richards (1980) who isolated 2 isoproteins of copper-rich metallothionein from the particulate fractions of human foetal liver, using buffers containing β-mercaptoethanol as extractants. Similarly, the insoluble copper containing granules in the livers of Bedlington terriers were also shown to contain copper-metallothionein (Johnson *et al*. 1981). The present study was designed to establish whether particulate forms of copper-metallothionein also accumulated in the liver of growing or adult animals under normal physiological conditions and to obtain information on the quantitative significance of this protein in pigs given copper supplemented diets.

METHODS

Animals

Twelve large white x (landrace x large white) pigs, aged about 6 weeks, were given a diet containing 300 mg copper/kg, for 8 weeks prior to slaughter. Livers were removed immediately at slaughter and stored at -20^0C until fractionated.

Analytical techniques

Details of the methods used for copper and zinc analysis, for estimation of thiol content, for polyacrylamide gel electrophoresis and for determining amino acid composition have been reported previously (Bremner & Young 1976).

Fractionation of livers

(i)   Samples (about 10 g) of liver were homogenized in 3 vols (v/w) of 10 mM Tris-acetate buffer, pH 7.4, centrifuged at 100 000 g for 1 hour, and the supernatant was fractionated on Sephadex G.75, as described previously. The pellet was then suspended in 1% β-mercaptoethanol (3.5 ml/g liver), freeze-thawed 3 times to aid disruption of membranes, and allowed to stand overnight at +4^{0}C. The mixture was then centrifuged and fractionated on Sephadex G.75 as above.

(ii)   In an alternative procedure, a sample (about 10 g) of liver was homogenized with 10 vols 10 mM Tris-acetate, pH 7.4, containing 1% β-mercaptoethanol and 1% sodium dodecyl sulphate (SDS). The homogenate was allowed to stand for 3 hours at 37^{0}C, whereupon it was centrifuged at 100 000 g and the supernatant fraction separated on Sephadex G 75.

Isolation of copper-metallothionein from particulate fractions

Samples (about 200 g) of liver were homogenized with 1.5 vols (v/w) of 10 mM Tris-acetate, pH 7.4, and centrifuged at 100 000 g for 1 hour. The pellet was homogenized with 1% β-mercaptoethanol (4 ml/g liver), freeze-thawed 3 times and allowed to stand at 37^{0}C for 3 hours. The mixture was centrifuged as above and the soluble extract concentrated 5-fold by ultrafiltration through Diaflo UM-2 membranes before fractionation on Sephadex G 75, using Tris-acetate containing 10 mM β-mercaptoethanol as eluant. The major copper-containing peak (with Ve/Vo = 2) was collected and further purified by ion-exchange chromatography on DEAE-Sephadex A 25, as described by Bremner and Young (1976). The fraction which did not bind to the DEAE-Sephadex and the 2 main fractions obtained on gradient elution of the column with 10-200 mM Tris-acetate containing β-mercaptoethanol were concentrated by ultrafiltration. They were further purified by gel filtration on Sephadex G 50, followed by affinity chromatography on Thiopropyl Sepharose 6B, as described by Ryden and Deutsch (1978). Flow rates were limited to only 2 ml/h during application of the sample and during elution of the copper-proteins with 50 mM β-mercaptoethanol in 50 mM Tris-HCl, pH 8.0.

RESULTS AND DISCUSSION

The pig livers contained 110.6 ± 12.3 μg copper and 76.0 ± 2.7 μg zinc/g wet weight, and about 46 and 77% of the tissue copper and zinc were recovered in the cytosol after the livers were homogenized in Tris-acetate buffer. Fractionation of the cytosol on Sephadex G 75 gave 3 copper- and zinc-containing fractions, the third of which, with Ve/Vo = 2, was metallothionein (Bremner & Young 1976). This contained 88 and 52% of the cytosolic copper and zinc respectively, the copper:zinc ratio in the protein being 1.4:1.   The remainder of the cytosolic copper was equally distributed between the high molecular weight fraction excluded by the Sephadex and that containing superoxide dismutase.

Subsequent extraction of the pellet with 1% β-mercaptoethanol removed a further 22 μg copper and 6.7 μg zinc/g liver. Fractionation of this extract on Sephadex G 75 appeared to give the same 3 copper-

containing fractions as were obtained from the cytosol and in a similar
ratio. Thus, 85% of the copper was present in the third fraction, with
Ve/Vo = 2, equivalent to about 17% of the total liver copper. The
copper content of this fraction was, however, much more than in the
cytosol, the copper:zinc ratio being 9:1.

Sequential extraction of the pig livers with 10 mM Tris-acetate and
with 1% β-mercaptoethanol therefore removed about 70% of the liver
copper and nearly 90% of this (63 μg/g liver) appeared to be present in
one fraction. Even more complete dissolution of hepatic copper was
achieved by extraction of the whole liver with a mixture of SDS and
mercaptoethanol at 37⁰C for 3 hours. Over 90% of the liver Cu was then
dissolved by this treatment, with 80% of it (77 μg copper/g liver)
present in the low molecular weight fraction separated by gel filtration.
The average copper:zinc ratio in this fraction was 2.3:1.

It seemed likely, in view of the previous characterization of copper-
metallothionein in the cytosol of pig liver (Bremner & Young 1976),
that the equivalent copper protein present in the particulate fractions
had the same composition. This was confirmed by fractionation of the
mercaptoethanol extract of the pellets obtained after centrifugation
from the Tris-acetate homogenates. The fractionation procedure used
was essentially the same as that used for the cytosolic protein (Bremner
& Young 1976), except that mercaptoethanol was included in all buffers
to overcome the oxidative changes that occur when metallothioneins of
high copper:zinc ratio are processed. In addition an affinity chromatography step had to be included to remove all contaminant protein.

Anion exchange chromatography on DEAE Sephadex A 25 of the major
fraction obtained on gel filtration of the mercaptoethanol extract gave
3 main sub-fractions, similar to those obtained from the cytosol
(Bremner & Young 1976). In most cases, about 50% of the copper did
not bind to the column and was eluted with the equilibrating buffer.
The other fractions (MT-I and MT-II) were eluted in about equal amounts
with 40 and 110 mM buffer. Total recovery of copper from the column
was about 85% and the isoproteins generally had copper:zinc ratios of
4-7:1.

Affinity chromatography of these 3 sub-fractions on Thiopropyl
Sepharose 6B gave proteins that were homogeneous and had amino acid
compositions typical of the cytosolic copper-metallothionein (Table 1).
The proteins contained only copper as bound metal, as zinc was
selectively removed during the affinity chromatography step. However,
if allowance was made for this zinc loss, the cysteine:metal ratio in
the proteins was about 2.7:1. The thiol:metal ratio in the proteins
was only 0.6:1, compared with about 2:1 for the cytosolic proteins,
but this is almost certainly a consequence of the removal of zinc and
subsequent formation of disulphide bonds.

These results show clearly that much of the copper in the non-
cytosolic fractions of pig liver is present as copper-metallothionein.
Including the cytosolic protein, up to 70% of the total hepatic copper
occurred in this form, indicating that metallothionein is the major
copper reserve in these livers. The association of copper in the
particulate fractions was therefore similar to that in the livers of

L.I.H.E.
THE MARKLAND LIBRARY

the human foetus and of Bedlington terriers (Riordan & Richards 1980; Johnson *et al.* 1981). However, copper does not invariably accumulate in this form, as no particulate copper-metallothionein was found in the livers of copper loaded sheep and rats (Mehra & Bremner unpublished results).

TABLE 1   Amino acid composition of isoproteins of copper-metallothioneins isolated after extraction of the particulate fraction of pig liver with mercaptoethanol solution

Amino acid	MT-I	MT-II	'Unbound' MT	Cytosolic CuMT-I
Cysteine	31.7	31.5	30.8	33.1
Aspartic acid	6.9	7.4	5.7	7.5
Methionine	2.7	2.1	5.0	1.7
Threonine	5.5	6.1	6.0	4.6
Serine	13.5	13.4	13.1	13.1
Glutamic acid	2.2	2.5	3.1	2.0
Glycine	12.2	12.5	12.0	11.2
Alanine	11.6	12.0	10.7	11.7
Valine	1.5	1.7	2.3	0.9
Isoleucine	1.8	2.1	1.7	1.5
Leucine	0.8	1.1	1.5	-
Lysine	10.3	9.5	10.7	11.4
Arginine	-	-	-	1.4

The removal of the particulate copper-metallothionein with mercaptoethanol or more efficiently with SDS/mercaptoethanol suggests that the particulate copper-protein may have originally contained disulphide bonds and been located in some membrane-bound form. The relatively low metal content of the protein (about 6%), even after allowance was made for zinc loss during affinity chromatography, is consistent with the original presence of some disulphide bonds. Although no direct information is available on the intracellular localization of the protein, it seems probable by analogy with other reports that it occurs mainly within the lysosomes.   As there were no major differences in the amino acid composition of the cytosolic and particulate forms of the protein, the lysosomal metallothionein may have arisen from sequestration of the cytosolic protein. This would be consistent with the resistance of copper-metallothionein to degradation by lysosomal enzymes *in vitro* (Bremner & Mehra 1983).

REFERENCES

BREMNER, I. 1979a. Copper toxicity studies using domestic and laboratory animals. In: *Copper in the environment - II*, edited by J.O. Nriagu, 285-306. New York: Wiley Interscience.

BREMNER, I. 1979b. Factors influencing the occurrence of copper thioneins in tissues. In: *Metallothionein*, edited by J.H.R. Kagi & M. Nordberg, 273-280. Basel: Birkhäuser-Verlag.

BREMNER, I. & MEHRA, R.K. 1983. Metallothionein: some aspects of its structure and function with special regard to its involvement in copper and zinc metabolism. *Chem. Sci.*, 21, 117-121.

BREMNER, I. & YOUNG, B.W. 1976. Isolation of copper, zinc-thioneins from pig liver. *Biochem. J.*, 155, 631-635.

JOHNSON, E.G., MORELL, A.G., STOCKERT, R.J. & STERNLIEB, I. 1981. Hepatic lysosomal copper protein in dogs with an inherited copper toxicosis. *Hepatology*, 1, 243-248.

PORTER, H. 1971. Neonatal hepatic mitochondrocuprein IV: sulfitolysis of the cystine-rich crude copper protein and isolation of a peptide containing more than 35% half-cystine. *Biochim. biophys. Acta*, 229, 143-154.

RIORDAN, J.R. & RICHARDS, V. 1980. Human fetal liver contains both zinc and copper rich forms of metallothionein. *J. biol. Chem.*, 255, 5380-5383.

RYDEN, L. & DEUTSCH, H.F. 1978. Preparation and properties of the major copper binding component in human fetal liver; its identification as metallothionein. *J. biol. Chem.*, 253, 519-524.

PROTON NMR STUDIES OF PLAICE LIVER METALLOTHIONEIN: METAL REMOVAL BY EDTA

J K NICHOLSON*, P J SADLER* and J OVERNELL**

*Department of Chemistry, Birkbeck College, University of London WC1E 7HX
**Institute of Marine Biochemistry, St Fitticks Road, Aberdeen AB1 3RA

ABSTRACT

Native Cd, Zn-metallothionein (MT) from plaice liver has been studied by (400 MHz) ^{1}H nuclear magnetic resonance (NMR) spectroscopy. When EDTA is added to the MT solution, removal of Zn and Cd from the protein can be directly monitored from the intensities of the shifted proton resonances due to the EDTA complexes of the 2 metals. Using this approach, which can be applied to other metalloproteins, both the amount and the kinetics of removal of zinc and cadmium can be studied for each metal separately, and data are also obtained on changes in protein tertiary structure due to metal removal. It was found that, when excess EDTA was added to plaice MT containing cadmium and zinc in a ratio of 10:1, almost all of the zinc but very little of the cadmium was removed from the protein over a period of 400 minutes at 298^0K. The total amount of metal (cadmium + zinc) extracted by EDTA was slightly less than one-seventh of the metal originally bound to the protein, suggesting that only *one* of the 7 postulated metal binding sites is involved. The removal of the other metals may be limited by kinetic or more probably thermodynamic factors. Significant changes in the chemical shifts of the methyl resonances of N-acetyl methionine were noted when metal was removed, suggesting that the EDTA-accessible metal binding site is close to the N terminal end of the protein.

INTRODUCTION

Metallothioneins are ubiquitous proteins containing about 61 amino acids of which 20 are cysteine residues, and are thought to bind a total of 7 Zn^{2+}, Cd^{2+} and sometimes Cu^+ ions depending on from which species and which organs the proteins are isolated (Webb 1979). The distribution of metal ions amongst the binding sites and their accessibility are the subjects of much current interest, because metallothioneins have been implicated as possible detoxification agents for certain potentially toxic metals such as cadmium and mercury, and as possible stores and homeostatic regulators of biochemically essential metal ions such as zinc and copper (see Cain (p 55) for a review of the general properties of metallothioneins).

Proton NMR is a powerful tool in the elucidation of protein tertiary structure in solution and is particularly informative about molecules with relatively low molecular weights such as metallothionein (7000 daltons). We have studied solutions of the native plaice liver metallothionein previously characterised by Overnell and Coombs (1979), both by conventional single pulse ^{1}H NMR spectroscopy and spin-echo Fourier transform (SEFT) ^{1}H NMR spectroscopy (which gives information about the most structurally mobile regions of the protein molecule) in the absence and in the presence of EDTA, a strong chelating agent for both zinc and cadmium. The non-destructive nature of NMR allows the removal of metal from protein to be monitored by observation of the separate ^{1}H resonances due to the stable Zn- and Cd-EDTA complexes, the intensities of which are a measure of the amount of metal removed from the protein in a given time. In addition, this approach allows simultaneous observations of changes in the proton NMR spectrum of the protein itself. These are related to changes in protein tertiary structure.

METHODS

1. Preparation of metallothionein: MT was isolated from the livers of cadmium treated plaice as previously described by Overnell and Coombs (1979). Tris buffer was removed from the protein solution by standard desalting methods. The metallothionein contained cadmium and zinc in a molar ratio of 10:1.

2. NMR measurements: ^{1}H NMR spectra were recorded in a Bruker WH 400 spectrometer operating at 400 MHz (9.4 Tesla) at an ambient temperature of 298^0K. A solution of 1.5 mg of plaice liver metallothionein in 0.5 ml of D_2O was prepared (ie approx 0.5 mM in protein) and adjusted to pH 8 with NaOD. The solution was transferred to a 5 mm O.D NMR tube in which measurements were made. Spin-echo spectra were recorded using 300 repetitions of the Hahn spin-echo pulse sequence (90^0-τ-180^0-τ-collect) with τ = 60 μs using a 9.5 μs 90^0 and a 19 μs 180^0 pulse. After standard single pulse and spin-echo spectra had been obtained, 25 μl of 100 mM EDTA in D_2O at pH 8.3 was added to the NMR tube giving a final concentration of 4.76 mM EDTA and approximately a 10-fold excess over the protein concentration. Starting immediately, consecutive spin-echo spectra were obtained at 15 minute intervals over a period of 400 minutes, each spectrum taking 15 minutes to run. At the end of the experiment, an aliquot of 200 μl of 20 mM $ZnSO_4$ was added to the solution (increasing the zinc concentration to 0.73 mM) and a further spectrum obtained to calibrate the Zn-EDTA concentration scale. This procedure was then repeated after a standard addition of $CdSO_4$.

RESULTS AND DISCUSSION

A variety of metal ions form stable complexes with EDTA at neutral pH. These include zinc (log K_{stab}=16) and cadmium (log K_{stab}=16.5): 2 of the metals commonly associated with metallothioneins in biological systems. We have used proton NMR to estimate the amounts of zinc and cadmium that are released by plaice liver metallothionein at various times after the protein was mixed with an excess of EDTA. The chelating agent itself has a simple proton NMR spectrum of 2 single lines with an intensity ratio of 2:1 at 3.64 and 3.25 ppm corresponding to the 'acetate' (4x-CH_2-COOH) and the 'ethylenic' (N-CH_2-CH_2-N) protons respectively. In the Zn- and Cd-EDTA complexes, the 'acetate'-CH_2-COOH protons are non-equivalent and give AB patterns at 3.45 ppm and 3.20 ppm. The 4 ethylenic (N-CH$_2$-CH$_2$-N) protons remained magnetically equivalent but shifted to 2.87 and 2.70 ppm for zinc and cadmium complexes respectively. The rate of exchange of EDTA between its metal bound and free forms is slow on the NMR timescale at 400 MHz ^{1}H frequency. The concentrations of the metal-EDTA complexes are proportional to the areas of their proton resonance peaks. Thus, the intensities of the sharp singlets at 2.87 (Zn) and 2.76 (Cd) ppm were used as a measure of the amount of each metal released at various times after the addition of a 10-fold excess of EDTA to the plaice liver metallothionein solution. At the end of the experiment, the metal-EDTA peak intensities were calibrated by the addition of a standard aliquot of each metal. We have used a similar NMR technique involving EDTA complexation to measure the concentrations of magnesium and calcium in plasma samples and cell extracts (3). This procedure can be extended to the measurement of a number of other diamagnetic metal ions which form complexes with EDTA.

Figure 1A shows a typical normal proton NMR spectrum of plaice liver metallothionein. A number of well-resolved resonances are seen which correspond with the known amino-acid composition of the protein. Figure 1B shows a proton NMR spectrum obtained by use of a Hahn spin-echo pulse sequence (90^0-τ-180^0-τ collect) with τ = 60 μs. This type of experiment eliminates broad components from the spectrum. These components are due to protons with rapid T$_2$ relaxation which are usually in relatively immobile parts of the protein. Thus, only the proton resonances from the most freely mobile regions of the protein are observed and this allows the proton resonances of Zn- and Cd-EDTA to be seen more clearly as in Figure 1C; with this τ value, singlets and triplets remain upright but doublets are inverted. A small peak was also observed at 2.56 ppm (superimposed upon the γ CH proton resonance of the glutamic acid residues) due to a calcium EDTA complex from a small calcium impurity in the protein. A typical plot of the rate of formation of Zn-EDTA and Cd-EDTA complexes with time after the addition of EDTA (5 mM) to a 0.5 mM solution of plaice liver MT is shown in Figure 1D. There is an initial rapid phase of metal loss from the protein; in the first few minutes of the reaction, more cadium is released than zinc. However, by T = 30 minutes, more zinc had been released from the protein. By the end of the experiment (T = 400 minutes), the concentration of Zn-EDTA was 0.28 mM, whereas Cd-EDTA was only 0.16 mM. This is an interesting result as the native protein contained considerably more cadmium than zinc (ratio 10:1). The total released metal (0.44 mM) was slightly less than one-seventh of the initial total protein bound metal concentration, perhaps suggesting that only one of the 7 metal ions bound to the protein is involved in reactions with EDTA, irrespective of its occupancy by zinc or cadmium. Other experiments of this type have shown that up to two-sevenths of the total metal can be released from plaice MT which contain higher amounts of zinc, and that the kinetics of metal removal vary according to the initial Cd:Zn ratio in the native protein. This finding probably reflects differences in the metal occupancy of the binding sites and different reactivities of individual sites with EDTA. It is important to note in this respect that the affinities of zinc and cadmium for EDTA are very similar, and therefore metal removal rates represent differences in the reactivity of the bound metals. Metallothioneins appear to contain 2 clusters (A and B) of 4 and 3 metal ions respectively

(Otvos & Armitage 1980). The metal ions which are removed here would appear to arise from the 3 cluster. Rat liver MT becomes susceptible to proteolytic cleavage after removal of zinc from the B cluster with EDTA (Winge & Miklossy 1982).

A marked change in the proton NMR spectrum of the protein itself occurred after the addition of EDTA and the removal of some of the metal. Particularly striking was a splitting of the singlet peaks assigned to the $S-CH_3$ and CH_3CO (acetyl) protons of the methionine residue. In particular, the $S-CH_3$ peak showed a marked splitting and chemical shift (0.0s ppm down-field of its normal position), that of the N acetyl methyl being much smaller (0.0069 ppm) (see Figures 1B and 1C). This is an intriguing result as previous workers have remarked on "The total invariance in chemical shift of the methionyl $S-CH_3$ protons in the apoprotein and in the holoprotein" (Vašák *et al.* 1980). The most likely explanation of this $S-CH_3$ shift is that the metal which is removed by EDTA comes from a binding site which is very close to the methionine residue in the protein. It was even considered by Sadler *et al.* (1978) that the terminal methionine S might act as an additional ligand in one of the cadmium binding sites in rat liver metallothionein. However, we cannot rule out the possibility that plaice MT is a mixture of 2 isoproteins which have yet to be separated. If the shift of the $S-CH_3$ resonance does indicate removal of metal from near the N terminal region of the protein, then this binding site could be the most accessible to EDTA and therefore partially account for its reactivity. Subtle changes in other protein proton resonances (see Figures 1B and 1C) were observed on metal removal, particularly those of the threonyl residues. A full analysis of these spectral changes is being made.

These experiments show that certain metal binding sites on metallothionein are particularly reactive and accessible to small ligands such as EDTA. This may be reflected in different physiological availabilities of the metals to exchange within cells, which is of particular interest in view of the possible functions of MT in essential metal homeostasis and the possibility that MT acts as an essential metal (Zn^{2+} or Cu^+) donor to apo-metalloenzymes (Beltramini & Lerch 1982).

The ^1H NMR techniques described here give a wealth of useful information about metal binding site reactivity and conformational changes in metalloproteins undergoing de-metallation, and promise to be of great value in the study of metallothionein biochemistry and that of metalloproteins in general.

ACKNOWLEDGEMENTS

We thank RTZ Services, MRC, SERC, NERC, ULIRS and the Nuffield Foundation for financial support, Heather Robbins for technical assistance and Dr Milan Vašák for useful discussion.

Figure 1 Effects of EDTA on plaice-MT

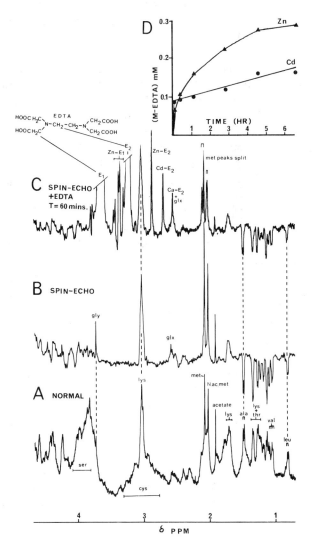

REFERENCES

BELTRAMINI, M. & LERCH, K. 1982. *Febs Lett.*, 142, 219-222.

NICHOLSON, J.K., BUCKINGHAM, M.J. & SADLER, P.J. 1983. High resolutions ^{1}H NMR of vertebrate blood and plasma. *Biochem. J.*, 211, 605-615.

OTVOS, J.D. & ARMITAGE, I.M. 1980. Structure of the metal clusters in rabbit liver metallothionein. *Proc. natn. Acad. Sci. USA*, 77, 7094-7098.

OVERNELL, J. & COOMBS, T.L. 1979. Purification and properties of plaice metallothionein, a cadmium binding protein from the liver of the plaice (*Pleuronectes platessa*). *Biochem. J.*, 183, 277-283.

SADLER, P.J., BAKKA, A. & BEYNON, P.J. 1978. [113]Cd nuclear magnetic resonance of metallothionein. *Febs Lett.*, 94, 315-318.

VAŠÁK, M., GALDES, A., HILLS, H.A.O., KAGI, J.H.R., BREMNER, I. & YOUNG, B.W. 1980. Investigations of the structure of metallothioneins by proton nuclear magnetic resonance spectroscopy. *Biochem.*, 19, 416-425.

WEBB, M. 1979. The metallothioneins. In: *The chemistry, biochemistry and biology of cadmium.* Amsterdam: Elsevier/North Holland.

WINGE, D.R. & MIKLOSSY, K.A. 1982. Domain nature of metallothionein. *J. Biol. Chem.*, 257, 3471-3476.

LIST OF CONTRIBUTORS AND PARTICIPANTS

S Andrews
Dept of Biology
Sunderland Polytechnic
Sunderland
SR1 3SD

M Birkhead
Edward Grey Institute
Dept of Zoology
South Parks Road
Oxford
OX1 3PS

K Cain
Toxicology Unit
Medical Research Council Laboratories
Woodmansterne Road
Carshalton
Surrey
SM5 4ES

A S Cooke
Nature Conservancy Council
Godwin House
George Street
Huntingdon
PE18 6BU

J A Cooke
Dept of Biology
Sunderland Polytechnic
Sunderland
SR1 3SD

T L Coombs
Institute of Marine Biochemistry
St Fitticks Road
Aberdeen
AB1 3RA

B Hunter
Dept of Zoology
University of Liverpool
PO Box 147
Liverpool
L69 3BX

M Hutton
Monitoring and Assessment Research Centre
The Octagon Building
Chelsea College
459A Fulham Road
London SW10 OQX

M Johnson
Dept of Botany
University of Liverpool
PO Box 147
Liverpool
L69 3BX

M D Kendall
Dept of Anatomy
St Thomas's Hospital
Medical School
London SE1 7EH

R K Mehra
Nutritional Biochemistry Dept
Rowett Research Institute
Greenburn Road
Bucksburn
Aberdeen
AB2 9SB

J K Nicholson
Dept of Chemistry
Birkbeck College
Malet Street
London WC1E 7HX

J Overnell
Institute of Marine Biochemistry
St Fitticks Road
Aberdeen
AB1 3RA

C Perrins
Edward Grey Institute
Dept of Zoology
South Parks Road
Oxford
OX1 3PS

B Pirie
Institute of Marine Biochemistry
St Fitticks Road
Aberdeen
AB1 3RA

P J Sadler
Dept of Chemistry
Birkbeck College
Malet Street
London WC1E 7HX

T Stowe
Royal Society for the Protection
  of Birds
The Lodge
Sandy
Bedfordshire
SG19 2DL

G Westlake
Environmental Chemistry Section
Tolworth Laboratory
MAFF
Government Buildings
Hook Rise South
Surbiton
Surrey
KT6 7NF

The following staff of the Institute were present:

D Osborn, W J Young, K R Bull, M C French, F Moriarty and P Freestone
Institute of Terrestrial Ecology, Monks Wood Experimental Station
Abbots Ripton, Huntingdon, Cambs PE17 2LS

388 P

130212

LINKED

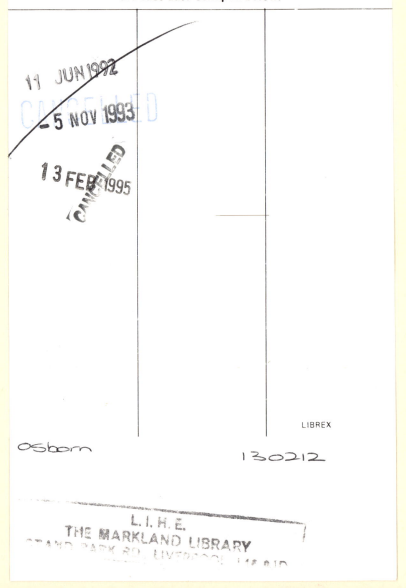

**This book is to be returned on or before
the last date stamped below.**

11 JUN 1992

CANCELLED

-5 NOV 1993

13 FEB 1995

LIBREX

Osborn   130212

L. I. H. E.
THE MARKLAND LIBRARY
STAND PARK RD. LIVERPOOL L16 9JD